The Moon

In Focus

Signs, Nodes and Eclipses

The Moon

In Focus

Signs, Nodes and Eclipses

Sasha Fenton

ZAMBEZI PUBLISHING LTD
www.zampub.com

First published in 2016 in the UK by Zambezi Publishing Ltd
Plymouth, Devon PL2 2EQ
Tel: +44 (0)1752 367 300   Fax: +44 (0)1752 350 453
email: info@zampub.com   www.zampub.com

Text copyright © 2016 Sasha Fenton
Sasha Fenton has asserted her moral right
to be identified as the author of this work
in accordance with sections 77 & 78 of
The Copyright, Designs and Patents Act 1988

British Library Cataloguing in Publication Data:
A catalogue record for this book is available from the British Library
ISBN: 978-1-903065-80-8
Illustrations copyright © 2016 Jan Budkowski,
Adobe Stock Images and others
Typesetting by Zambezi Publishing Ltd, Plymouth
Printed in the UK by Lightning Source UK

About the Author

Sasha Fenton became a professional astrologer, palmist and Tarot card reader in 1974, but wound down her consultancy when her writing took off. She has written 129 non-fiction books, and a ten years' worth of chapters for the Llewellyn Sun Sign books. Her sales now approach seven million copies, with translations into fifteen languages. In 2013, Sasha took to writing fiction and she is now producing a series of "Tudorland" novels, of which "Sophie's Inheritance" and "Lucy's Dilemma" are the first.

Having written stars columns for many papers and magazines over the years, including Woman's Own and the Sunday People, Sasha has also had her own radio and television programmes and has broadcast for many UK radio and television stations, as well as several in the USA, Australia and South Africa. She has taught, lectured and broadcast all over the world, including at the prestigious Mind, Body and Spirit Festivals in London, Sydney and Melbourne and in Johannesburg and Cape Town.

Sasha has been President of the "British Astrological and Psychic Society" (BAPS), Chair for the "Advisory Panel on Astrological Education", and a member of the Executive Council of the "Writers' Guild of Great Britain".

Sasha and Jan run Zambezi Publishing Limited and Stellium Ltd, producing books and ebooks of many different kinds. She is married to Jan Budkowski, and has two children and four grandchildren.

Contents

Chapter One:
The Moon and
the Jargon

The moon fascinates those who are interested in astrology, because it reveals our inner personality and our true desires, needs and ambitions in a way that the more external sun and rising signs cannot. The moon rules our feelings, emotions, most heartfelt requirements and real motivations. The moon's sign and house can explain far more about our real agenda than can ever be obvious from first, second and often even later impressions. However, there is much more to the moon in astrology than just the sign and house that it occupies. This book covers the moon's signs and houses, but it also deals with several other fascinating topics and features, such as the nodes of the moon, eclipses and occultations. It includes full interpretations of what these mean in terms of character, karma and events. Nothing in this book is beyond a complete beginner, but it also tackles topics that skilled astrologers rarely consider.

Every job, hobby and interest has its own jargon, which is often a form of shorthand for something that would otherwise require too many words. Astrology has plenty of its own jargon, but once you know what the words mean and the techniques that they describe, it's easy to understand. This book demystifies the jargon and gives full interpretations of all the techniques, thus allowing you to get started on them right away.

This book includes all the necessary data for the methods that are described, but if you want complete accuracy, you will need to take things a step further. Here are several options:

- Nowadays, there are many websites, such as www.equinoxastrology.com that offer a large range of charts, interpretations and many other astrological services; some for free, some not.
- There are even astrology apps for smartphones; check out your app store.
- You can visit a good astrologer.
- You can find other chart services in astrology magazines. Such services are also often advertised in the back of many ordinary

women's magazines. Some of these are better than others, though.

- You can purchase inexpensive software and make up your own chart. Such software now can even include explanations and interpretations.

- If you are considering serious astrology rather than simply doing a few charts for interest's sake, you will need to buy the kind of software that the professionals use. Any astrology magazine will carry advertisements for this, an online search will show you others, and it's worth sending off to all the companies for their brochures. At present, I use a few good programs for different purposes, of which my preferred one is called Solar Fire. You should always investigate software features to establish for yourself which program is likely to meet your specific requirements.

- You can take a course in basic astrology and learn the subject in full.

There are some tables in this book to make life easier, but larger tables, such as the moon through the zodiac signs, aren't included. However, the information is available online; try a search for this information if you don't have an astrology program of your own.

So, now let us look at a brief overview of the topics covered in this book. Each topic is fully explained later on.

The Moon's Signs and Houses
You can discover where the moon was when you were born, including the sign and the house that it occupied. This explains what the signs and houses are and gives full interpretations.

The Nodes of the Moon
The moon's nodes are the points where the path of the moon crosses the plane of the path of the sun. The sun's path is called the ecliptic. The north node is the point where the moon crosses

the ecliptic in a northerly direction and the south node is the point where the moon crosses it fourteen days later in a southerly direction. The position that these nodes were in on the day of your birth describes not only aspects of your character, but also the karmic issues that you will have to face during your lifetime.

Eclipses of the Moon

I am sure that everybody knows what an eclipse is, but few people realize that there are an average of four each year. The majority of these are partial eclipses, but as far as astrology is concerned, an eclipse is a powerful event, whether partial or full. The chapter on eclipses explains why the period of time around an eclipse can be difficult, and how it's likely to manifest itself in each sign.

Occultations by the Moon

Only a handful of astrologers even consider occultations, which are simply those occasions when the moon passes directly in front of a planet. This is a pity because occultations are easy to plot, and their effect on a person's life is powerful and immediate. An occultation can explain why someone goes through a difficult patch, even when there is nothing else showing up on the chart to account for it.

Summing Things Up

This book is aimed at those who are not yet into astrology, those who are getting into it and those who already know quite a bit about it. It's more involved and far more interesting than a basic sun or moon sign book, but it doesn't take an Einstein to understand or use the systems described in it. The book contains a simple method for finding a person's moon sign and house; it gives lists of dates for the positions of the moon's nodes and for eclipses. The occultations are harder to list but the book shows how to go about finding these. There is nothing that an absolute beginner cannot work out for himself. While this book will get a beginner up and running, there is no doubt that you will soon want complete accuracy, down to the exact degree and minute of a sign and house. It's likely that you will use this book as a starting point and then move on to higher levels of astrology.

Chapter Two:
The Moon

The moon is about a quarter of a million miles away from the earth and its surface area is about the size of Asia, although its highest mountains are higher than any on earth. It takes 27.32 days for the moon to circle the earth, but both the moon and the earth orbit the sun, which means that it takes an average of 28 days for the moon to move through all twelve signs of the zodiac. The moon is slowly moving away from the earth. Some say that in past millennia the moon shared some of the earth's gravity, which may have accounted for the ability of such huge animals as dinosaurs to exist.

The moon does not shed any light of its own; its surface only reflects the light of the sun. The phases of the moon depend upon its position during the course of each orbit. When it's between the earth and the sun the moon is dark for a short while, then as light begins to reach the surface of the moon once more, a sliver of new moon can be seen. When the moon is on the far side of the earth to the sun, the whole of its face is lit up, and this is a full moon.

Astrologers call the sun and moon planets for the sake of convenience, even though we are perfectly well aware that they are not. The moon is the closest planet to the earth and it moves through the zodiac very rapidly, spending an average of two-and-a-half days in each sign. Compare this with the sun, which takes a month to run through a sign. This means that a person who was born only a few hours later than you could well have the moon in the next sign and house along.

Your Secret Moon

The theory is that the sun rules your outer personality while the moon rules your emotions and feelings. The moon rules the kind of inner drives that may be hidden from others - at least upon first acquaintance. The moon rules your emotions, how they change, your changing mannerisms and the different phases of your moods and feelings. It rules your habits and your behaviour when you are not entirely in control of yourself - perhaps when you are sick, depressed, madly in love or under the influence of alcohol.

Having said this, many people behave far more like their moon sign when they are young and only develop their sun sign personality later in life. This makes sense when one considers that the sun is a dynamic planet that talks about worldly success, status and achievement - as opposed to the moon, which rules natural, habitual, instinctive, unthinking and intuitive forms of behaviour. It takes time to become a pillar of the community, while we are born with our instincts and intuition firmly in place.

The moon rules reactions rather than actions. If you choose to take a day off or to fix the car rather than bake a cake, that's a matter of choice. Now consider the situation when you get up late, the weather is bad and everything you try to do is delayed or frustrated. You arrive late for work and the boss is looking for you, someone has spilled coffee on your paperwork and your favourite pen has gone missing from your desk. None of this is earth shattering, but the way you cope with it and your specific reaction will depend upon the natal position of your moon, the sign the transiting moon is in on this particular day. Conditions such as a new or full moon is full, an eclipse or an occultation will also have a bearing.

Nurturing and Motherhood
The moon represents the female principle in your life and this may be a mother or mother figure, a female partner or wife, and women generally - although females in general are often also ruled by Venus. The moon is associated with the experience of being raised or nurtured, and thus your relationship with the person who did most for you when you were a child - regardless of whether this was your mother, father or someone entirely different. Thus, the moon refers to childhood experiences and the mark these left upon you - for good or ill.

Anatomically the moon rules the breasts, stomach, fluid balance of the body, the digestion, glandular secretions, the left eye of the male and the right eye of the female.

At Home and Abroad

The moon also rules a whole host of domestic matters. It's of particular importance in connection with the home and also to any property that you may buy, sell, rent, let out to others or deal with in any other way. The moon also rules the kitchen, refrigerator, cooker and everything to do with growing, buying, storing and using food. It rules cows and everything to do with milk - in olden days, it was said to rule milkmaids, but perhaps nowadays it rules those who produce and sell milk and milk products. It also rules beer and breweries.

Old time astrologers saw the moon as a fast and restless planet that vanished into the sea each month. Thus, it's said to rule travel, movement and restlessness. The moon is associated with rivers and the sea, thus also ships, sea voyages, fishermen, fishing nets and by extension sewing and sail-making. Much of this has now been passed over to the rulership of Neptune, but people with a strong moon on their charts still love to live near water.

Added Information for Those Who Are Further Into Astrology

If the moon is close to your ascendant, this will increase its influence and you may look more like your moon sign than your sun sign. In this case, or if the moon is in a considerably earlier house than your sun, you may behave and act according to your emotions at first, using logic and intelligence later. The chances are that your mother was a strong influence, either as a loving and caring figure and a wonderful role-model, or in a damaging and unloving way.

The moon traditionally rules Cancer; it's exalted in Taurus, in detriment in Capricorn and in fall in Scorpio. The glyph representing the moon in astrology is the moon in the first quarter.

How to Find Your Moon Sign

If you don't know which sign your moon was in when you were born, the simple tables in this chapter will do the trick. This system is very good, but, like all such generalized systems, it's not a hundred per cent accurate; for instance, if you were born with the moon "on the cusp", which means that the moon was close to the place where two signs meet on the day that you were born, the table may come up with the adjoining sign to yours. For the time being, your best bet is to use the table to find your moon sign and then to read the signs that come before and after, because the chances are that you will be able to spot yours when you read the through the character and lifestyle descriptions.

When you have some spare time, you can download one of the free astrology Apps for your smartphone, and find the complete planetary picture for your date, place and time of birth.

Working out Your Moon Sign

- Looking along the line that contains your year of birth in the Year / Month table that follows, note down the Sun sign underneath your month of birth, as shown on the right-hand side of the table.
- Find your day of birth in the smaller Day table, and note down the number beneath it.
- In the numbered list of astrological signs, find the sign you note down from the Month portion of the Year / Month table.
- From there, count onwards the number you found in the Day table, and your moon sign is the sign you reach.

Here is an example, using a birth date of February 9 1965:

- Looking at the Year / Month table, along the line containing 1965, under the month of February, we find the zodiac sign Cap (Capricorn).
- Under day 9 in the Day table, we note down the number 4.

- In the numbered zodiac sign list, we find that Capricorn is item 10.
- Counting four items on from there will take us through to Taurus. (i.e. After Capricorn, count Aquarius, Pisces, up to Aries, and the fourth item is Taurus).

So, in our example, the person's moon sign is Taurus.

Now, try this one, for Maria, who was born on November 9, 1070:

Nov, 1970 = Sag (Sagittarius)

Day no. 9 = plus 4

Sagittarius plus 4 = Aries

As it happens, Maria's moon is very late in Pisces, which shows that for this system, it's worth reading the moon sign on either side of the one that you calculate from these tables, to check which one is more appropriate.

Day Table

Exact Day of Birth							
1	2	3	4	5	6	7	8
0	1	1	1	2	2	3	3
9	10	11	12	13	14	15	16
4	4	5	5	5	6	6	7
17	18	19	20	21	22	23	24
7	8	8	9	9	10	10	10
25	26	27	28	29	30	31	
11	11	12	12	1	1	2	

YEAR / MONTH TABLE

						Jan	Feb	Mar	Apr	May	Jun	Jul	Aug	Sep	Oct	Nov	Dec
1920	1939	1958	1977	1996	2015	Tau	Can	Can	Vir	Lib	Sag	Cap	Aqu	Ari	Tau	Can	Leo
1921	1940	1959	1978	1997	2016	Lib	Sco	Sag	Cap	Aqu	Ari	Tau	Can	Leo	Vir	Sco	Sag
1922	1941	1960	1979	1998	2017	Aqu	Ari	Ari	Gem	Can	Leo	Vir	Sco	Cap	Aqu	Ari	Tau
1923	1942	1961	1980	1999	2018	Gem	Leo	Leo	Lib	Sco	Cap	Aqu	Ari	Tau	Gem	Leo	Vir
1924	1943	1962	1981	2000	2019	Sco	Sag	Cap	Aqu	Ari	Tau	Gem	Leo	Lib	Sco	Sag	Cap
1925	1944	1963	1982	2001	2020	Pis	Tau	Tau	Can	Leo	Lib	Sco	Sag	Aqu	Pis	Tau	Gem
1926	1945	1964	1983	2002	2021	Leo	Vir	Lib	Sco	Sag	Aqu	Pis	Tau	Can	Leo	Vir	Lib
1927	1946	1965	1984	2003	2022	Sag	Cap	Aqu	Pis	Tau	Gem	Leo	Vir	Sco	Sag	Aqu	Pis
1928	1947	1966	1985	2004	2023	Ari	Gem	Gem	Leo	Vir	Sco	Sag	Aqu	Pis	Ari	Gem	Can
1929	1948	1967	1986	2005	2024	Vir	Sco	Sco	Cap	Aqu	Pis	Tau	Gem	Leo	Vir	Lib	Sag
1930	1949	1968	1987	2006	2025	Cap	Pis	Pis	Tau	Gem	Leo	Vir	Sco	Sag	Cap	Pis	Ari
1031	1950	1969	1988	2007	2026	Tau	Can	Can	Vir	Lib	Sag	Cap	Aqu	Ari	Gem	Can	Leo
1932	1951	1970	1989	2008	2027	Lib	Sag	Sag	Aqu	Pis	Tau	Gem	Leo	Vir	Lib	Sag	Cap
1933	1952	1971	1990	2009	2028	Pis	Ari	Tau	Gem	Can	Vir	Lib	Sco	Cap	Aqu	Ari	Tau
1934	1953	1972	1991	2010	2029	Can	Vir	Vir	Lib	Sag	Cap	Pis	Tau	Gem	Can	Vir	Lib
1935	1954	1973	1992	2011	2030	Sco	Cap	Cap	Pis	Ari	Gem	Can	Vir	Sco	Sag	Cap	Aqu
1936	1955	1974	1993	2012	2031	Ari	Tau	Gem	Leo	Vir	Lib	Sco	Cap	Pis	Ari	Tau	Can
1937	1956	1975	1994	2013	2032	Leo	Lib	Lib	Sag	Cap	Pis	Ari	Tau	Can	Leo	Lib	Sco
1938	1957	1976	1995	2014	2033	Cap	Aqu	Pis	Ari	Tau	Can	Leo	Lib	Sco	Cap	Aqu	Ari

ZODIAC SIGNS	
1	Aries
2	Taurus
3	Gemini
4	Cancer
5	Leo
6	Virgo
7	Libra
8	Scorpio
9	Sagittarius
10	Capricorn
11	Aquarius
12	Pisces

Chapter Three:
The Zodiac

The signs of the zodiac are divided into three different types of group, these being gender, element and quality. In this chapter, we see how the elements and qualities affect the moon, the nodes and eclipses.

As you will see later, the nodes of the moon and lunar eclipse involve two signs, which are opposite to each other on the horoscope wheel. The following illustration shows all the signs, and it's easy to see each sign's opposite sign.

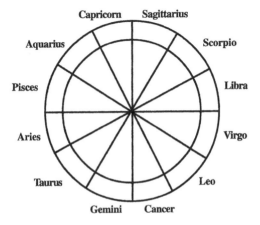

Each of the opposing signs shares the same Gender and Quality, but their Elements are different. Fire and Air are always linked, as are Earth and Water.

THE OPPOSING SIGNS

Aries *Masculine, fire, cardinal*	Libra *Masculine, air, cardinal*
Taurus *Feminine, earth, fixed*	Scorpio *Feminine, water, fixed*
Gemini *Masculine, air, mutable*	Sagittarius *Masculine, fire, mutable*
Cancer *Feminine, water, cardinal*	Capricorn Feminine, earth, cardinal
Leo *Masculine, fire, fixed*	Aquarius *Masculine, air, fixed*
Virgo *Feminine, earth, mutable*	Pisces *Feminine, water, mutable*
Libra *Masculine, air, cardinal*	Aries *Masculine, fire, cardinal*
Scorpio *Feminine, water, fixed*	Taurus *Feminine, earth, fixed*
Sagittarius *Masculine, fire, mutable*	Gemini *Masculine, air, mutable*
Capricorn *Feminine, earth, cardinal*	Cancer *Feminine, water, cardinal*
Aquarius *Masculine, air, fixed*	Leo *Masculine, fire, fixed*
Pisces *Feminine, water, mutable*	Virgo *Feminine, earth, mutable*

Gender

Every sign is either masculine or feminine, but this has nothing to do with sexuality. Some astrologers call these signs positive and negative and their energies are perhaps even better expressed as yang and yin. All the signs belonging to the fire and air elements are masculine or positive, while all the signs belonging to the earth and water elements are feminine or negative.

The Genders of the signs

The Elements

Each sign of the zodiac belongs to an element of fire, earth, air or water.

The Elements of the signs

The Elements

The Fire Group
Masculine
Aries, Leo, Sagittarius

This group is associated with speed, quick thinking and fast responses. People with the moon in fire signs don't lack initiative and they can get things off the ground quickly when they want to. They may lack patience with slower people, it doesn't take much to irritate or anger them and they don't stand fools gladly. These subjects can concentrate on a project that interest them: but they soon become bored by those that don't. They are generous and good hearted as long as others don't take advantage of their generosity. There are times when their self-centred attitude can make them appear arrogant, but they are not as confident as they look and they can be a prey to sudden fits of depression. They need a steady partner who they can rely upon.

The moon is especially associated with water signs, so it may not be all that comfortable when placed in this group. Sudden enthusiasms may not be all that easy to carry through and the person's changeable emotions may confuse those who are around him.

Childhood experiences may not be all that comfortable, as the father may be weak or a martinet, the mother demanding, and either or both parents may disappear off the scene at one time or another.

The Earth Group
Feminine
Taurus, Virgo, Capricorn

When the moon is in an earth sign, there is a certain amount of practicality. These people find it easy to cope with domestic life and daily decision-making. They are as good with their hands as they are with their heads. They tend to be thorough, careful and capable of dealing with details. They make very reliable friends and partners. There is a certain stubbornness here, which helps them to finish what they start, but on the other hand, this can tie them into a job, relationship or some other situation longer than is good for them.

Earth moon subjects fear poverty, so they may work hard for their money or hang on to what they have, to the point of becoming tight-fisted. They are particularly keen on family life and they will try to keep a family together come what may. Work is important to them and they may define themselves by their successes or failures at work. The moon is quite comfortable in earth signs, but this element can make it hard for a person to adapt to new circumstances.

As a child, this person would have had at least a reasonable relationship with his father - indeed, this may have been quite a successful one. The mother would have been a powerful and possibly repressive figure who made demands upon the child and some of her behaviour may have been so contradictory that it confused the child at times.

The Air Group
Masculine
Gemini, Libra, Aquarius
Air signs are mainly concerned with logic, a strong mind and intellect. This moon placement adds intelligence and speedy thinking, but it can make it difficult for these subjects to cope with emotional issues or with their own feelings. Some of these subjects avoid emotional situations by burying themselves in work or studies. Others find themselves at odds with partners who are controlled by feelings. These people can be unrealistic, with dreams that are far above what life can hope to offer them -

although some do make great achievements in their careers. Some are critical and arrogant.

Issues concerning saving and spending or giving and taking can be an issue for these people. They are inclined to be extremely self-indulgent, but they can be peculiar when it comes to dealing with others, being either over-generous or unnecessarily tight-fisted. Some give freely to one person while keeping others on a tight leash. Just like their confusing financial ideas, their emotions can be equally confusing, especially to themselves.

Childhood relationships are pretty reasonable and their parents would have done all they could to give these subjects a good education and plenty of intellectual stimulus, and the parents may well have been extremely bright themselves. Both parents were influential. The child was encouraged to think independently and to be independent at an early age, but whether this was due to good parenting or neglectful one depends upon specific circumstances.

The Water Group
Feminine
Cancer, Scorpio, Pisces
The moon is supposed to be most at home in water signs, but this can put more emphasis on the emotional and intuitive features in the personality than is comfortable. These people may become so awash with feelings and emotions that they find themselves overwhelmed by them at times. These subjects can switch from elation to depression in a short space of time, sometimes without quite knowing why. Part of the problem is that they are so quick to pick up on atmospheres, vibes and surrounding emotions that they absorb the feelings of those who are around them. These sensitive people love deeply, but they also hate with equal passion. Where practicalities such as money are concerned, they fear poverty but they may have to cope with this at some point in their lives.

The childhood can be difficult because these youngsters are not altogether on the same wavelength as those who are around them at home and at school.

The Qualities

Each sign of the zodiac belongs to a quality, and these are cardinal, fixed and mutable. These are not easy to define where it comes to the moon sign, because the energies are rather subtle.

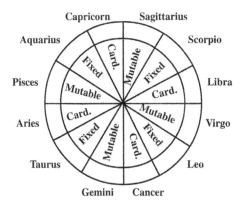

The Qualities of the signs

The Cardinal Group
Aries, Cancer, Libra, Capricorn
These people will try to maintain equilibrium in their lives, but their destiny tends to force them to cope with quite profound changes in circumstances at certain times. When this happens, they take the bull by the horns, take the kaleidoscope pieces of their lives and create a new pattern. They may move house several times, change partners or live through sudden bereavements or sudden career changes. They can think on their feet when they have to and they can be extremely successful and even wealthy at some points in

their lives. Sometimes this is due to such things as a lucky marriage or inheritance. Whether they stay rich or not depends upon other factors on their charts.

The Fixed Group

Taurus, Leo, Scorpio, Aquarius

People with the moon in fixed signs are stubborn. They find it difficult to see other ways of going about things or of solving problems, and they become disappointed and depressed if their plans don't work out. They go after what they want in a dogged manner, in some cases by hard work, sometimes by expecting others to provide for them. They can be self-absorbed, but they are usually reliable and very attached to their families and to those who they love. Some of these people are extremely emotional, others find it almost impossible to deal with emotions - even their own.

The Mutable Group

Gemini, Virgo, Sagittarius, Pisces

These people have an uncanny knack of attracting highly emotional people to them, and in some cases, they also attract lame ducks. They may sacrifice themselves for the sake of loved ones or even for strangers and they can exhaust themselves in the process. They need to become more discriminating in their choice of relatives, friends and colleagues and not to jump in and help others too quickly or take on more than they can reasonably cope with. Their own fluctuating moods make them difficult to understand, and they can suffer from chronic ailments if they don't get a grip on their lives. These are the easiest of all the quality types to get on with, but sometimes they are neurotic, nervy, emotional, quick to anger and difficult to live with.

Chapter Four:
The Moon through the Signs

This chapter looks at the moon through the signs of the zodiac. As explained at the beginning of this book, much of the information that you need is in here, but some data, such as your own moon sign, would take far too much space to put into a book; it's much simpler to find your own moon sign (or accurate sun sign, for that matter), by looking online.

The moon is associated with the cardinal water sign of Cancer, which means that it's particularly powerful in the water signs of Cancer, Scorpio and Pisces, and it can exert a strong influence in the cardinal signs of Aries, Cancer, Libra and Capricorn. It's likely to be of less importance in the other signs, as it can be overshadowed to some extent by the sun or other features on a horoscope. Traditional (ancient) forms of astrology tell us that the moon is exalted in Taurus, in detriment in Capricorn and in fall in Scorpio. It's true that its influence can be hard to live with in Capricorn and Scorpio - but, in Taurus, which should be a wonderful placement for it, it brings a mother figure into the picture who may be hard to handle.

The Moon in Aries

The element for Aries is fire and the quality is cardinal, so the moon is not particularly at home in this sign and this makes you react too quickly to stimuli. You may jump to conclusions prematurely or go after something with all guns blazing, only later discovering that your judgment was mistaken. Being quick to think, feel and act is fine, but you may find it hard to control your temper at times. You may over-dramatize your emotions and feelings, or your inner tension may erupt into sickness. Being competitive, you would do well to channel your energies into a creative enterprise, a career or a sporting outlet. You may have been a noisy and demanding child, which may not always have endeared you to those who took care of you. Being quick to learn and quick on the uptake, you probably did well at school - at least with those subjects that held your interest. You may have been good at sports, which would have been a positive outlet for your energies. Your biggest problem is that you soon become bored and find it hard to see projects through to the end. It's likely that once you reached your teens, you found yourself at odds with one of your parents (probably your father) and head to head screaming matches may have been the order of the day for a while.

Deep down inside, you want to deal fairly with others, but you will always need to assert your own feelings, regardless of whether you are right or wrong. You can be inconsistent in the way you behave because so much depends on your emotions at any given point in time. If you are on an emotional high, that's wonderful for everyone around you, but if you are down in the dumps, then others will be the first to feel the effects, possibly with no idea of why you are upset or what they have done to make you behave badly towards them.

You may be eager to form lasting relationships but lack the staying power that allows an encounter to develop into a permanent relationship. The demands and responsibilities of love can make you afraid to enter into a commitment or to risk rejection, and this means that you may be alone for longer than you want. You have a large circle of friends, acquaintances and business contacts, though few of these are truly close. Those who you want must make the first move and admit to their true feelings for you, before you can open up and take on a permanent relationship or commitment. If you become a parent, you will be loving and affectionate, but you may ask your children to do more than they are capable of. You may be a competitive or pushy parent or you may become somewhat bored by the long-term demands of parenthood.

It's important for you to maintain a well balanced diet and only to eat when you are calm and relaxed. If you eat when you are wound up or upset, you could suffer from headaches or other discomfort.

Although you have a wonderful imagination, which can be expressed creatively, you can also create imagined physical problems, so it's important for you to learn how to relax. Try painting or joining others in sports and games.

The Moon in Taurus

The element is earth and the quality fixed, and this makes you the most stubborn and determined off all the different types. You may be reliable and sensible, but the fluctuating and emotional nature of the moon can make you moody, changeable and quite unhappy within yourself at times. You need a stable life, a settled job and a reliable lover or life-partner, but you may find this difficult to achieve. Some moon in Taurus people insist on wanting people who don't really want them, and hanging on for years in the hopes that the other person will eventually choose them over whatever or whoever else is in the way. If you are the more sensible type, you will choose a reasonable partner and then become settled and happy.

You are responsible in family situations and personal relationships. You will do your best to earn the money that the family needs, and you can be relied upon in practical matters. However, you may be too obstinate and dictatorial for comfort and your tendency to see things your way and to refuse to back down or to allow others to express their feelings and to have their needs met can cause relationship breakdowns.

You need a good standard of living and you may go about achieving this by working hard in a relatively unexciting but steady kind of job or possibly by marrying a wealthy partner. You are an excellent home maker and true family person. The best way for you to relax, is to cook something nice for the family, to work around the house, take on do-it-yourself tasks, grow food and flowers and to listen to music or perhaps to play a musical instrument.

You take your duties as a parent very seriously and it would take an earthquake for you to abandon a child. However, your tendency to become depressed and to lose your temper can frighten or alienate a child. You mean well, but your moods can be hard for children to live with. The worst aspect of this moon is a tendency to escapism, possibly by losing yourself in alcohol, or, less dangerously, in dreams of romance and travel.

The Moon in Gemini

The element for Gemini is air and the quality mutable so this is a reasonably comfortable position for the moon to find itself in. The mutability of this sign and the restless nature of the moon make take you away from your home area, perhaps even to another country at some point. You may move house frequently or learn to adapt to new people and places, whether you want to or not.

You are blessed with a quick mind and a great sense of humour which stands you in good stead when life is difficult. Despite the fact that this is a mutable sign, you are apt to find a job and stick to it. You may take your time about getting into relationships, but once you do, you are steadfast and you will try to stick out a partnership even when it's not a particularly good one. Friendships are less important to you than family relationships. There is an element of restlessness that makes you choose a job that offers a variety of tasks or that puts you in touch with many people during the course of the day. You are an excellent communicator, either on the phone or on paper and you are always well-informed and up-to-date. Your fund of knowledge and talent for communications may take you into the travel trade, journalism or teaching. You all benefit from short breaks and changes of scene.

Your mind is organized and logical, but you may find it hard to understand, handle or even admit to your feelings and emotions, although you can be uncannily intuitive about other people. Some of you attach yourself to a partner who is emotional and moody, which gives you an opportunity to learn about emotions and moodiness at second hand. If you become downhearted, you become nervy and sick, so you must guard against becoming over-wrought or worn out by life by taking short breaks and perhaps by using aromatherapy and massages.

Your parents will have been reasonable and they will have done all they can to help you get on at school and so on, but they may not have been able to help as much as they would like, possibly due to their own lack of education. Your love of reading will have helped you to overcome any shortage of information. As a parent yourself, you will do all you can to help your children receive the intellectual stimulus and love that they need.

The Moon in Cancer

The moon is in its own sign here, so the water element and cardinal quality of the sign is not at odds with this planet. However, it can make you over-sensitive and prone to pick up on atmospheres and the feelings of others.

You have a kind heart and an extremely sensitive nature, you are a sympathetic listener and you do all you can for those who need a hand. However, you can be moody and extremely sarcastic when someone hurts you - and sometimes even when no hurt is intended. You appreciate consideration from others in relation-ships and a steady and reliable partner. You may go through life with unresolved issues surrounding your own mother or you may see her as a strong woman and an excellent role model. You are fairly studious and you have a very good memory which helps you to gather and retain information, but your long memory means that you carry past hurts around for too long. You are steadfast and reliable in relationships and you care deeply for your own children. Your protective instincts may take you into teaching or one of the caring professions such as looking after young children, the elderly or disadvantaged people of some kind. Alternatively, you may choose to run a guesthouse or a small business that provides the public with something that it needs. Working from home appeals to you.

You need to relax by walking along a shore line or taking holidays on water. Reading, art, dressmaking, gardening, fixing the car, decorating and cooking are all relaxing and creative occupations for those times when you need to switch off.

The Moon in Leo

Leo is a fixed fire sign which is not a particularly comfortable placement for the moon, so you may sometimes feel stifled by the situation you find yourself living in. Your feelings are strong and you need to express them, and you are quick to reach out to others for love or for friendship. Your emotions are fairly close to the surface and you may dramatize them at times, because there is a strong element of self-centredness here.

Wanting the best things in life, you will try to earn the money that will furnish you with the kind of home or lifestyle that you need. You need good food and drink, plenty of books and gadgets and enough of everything for yourself and for those who you love. Some people with this moon placement behave in a totally the opposite manner to this, in that they have very little pride and don't want to make an effort in life, although they still want a nice lifestyle, so they marry someone who can provide this for them. Some of you are the souls of generosity, others are extremely miserly.

There is an element of vanity and fussiness here which can manifest itself in fussing over your clothes, appearance and especially your hair. You may also seek to be the centre of everyone else's attention or so self-indulgent that nobody else gets a look in. Alternatively, you can be very loving and able to create a close and supportive partnership and far more indulgent to those you love than you are to yourself.

Your own parents were probably quite reasonable, but there is a hint that you go on to choose a very different lifestyle to theirs. Sometimes there are issues of class, culture and status, in which you move up or down the class scale or hold different values from those of your parents. Although extremely loving, caring and responsible as a parent, you may be a little too demanding for comfort, although you mean well. You enjoy an opportunity to show your children off and to be proud of them, but this can make you a little too competitive on their behalf, expecting them to achieve all kinds of scholastic and other goals that may be beyond them.

You can relax by getting involved in any creative activity or by enjoying the company of friends, playing with children or with your pets.

The Moon in Virgo

The mutable earth element of Virgo should fit nicely with the moon, but it seems to create a tension between the desire for the emotions to flow like water, and the practicality of this particular earth sign. You may be able to change many things about your life, but it's hard for you to change the way you feel.

Your methodical and painstaking emotional nature may mean that you find it difficult to enjoy casual relationships. You examine people too closely and discover flaws that you cannot tolerate. If a person passes your test, you may only enjoy them for what they do and fail to appreciate them for who they are. You are right to regard yourself as an analyst, but it's also very important to be humane when passing judgement on others because the perfection you search for is unlikely to be present. You may have such a strong expectation of rejection by others that when someone appears to want you, you make impossible demands that are unconsciously designed to push their patience to the limit. Unless someone is able to establish their true worth to you, you don't waste your time on them. The biggest danger in any relationship for you is to overdo the criticism, though when you are able to accept others with all their faults and frailties, your life will gain dimension and interest. You may show your love for others by doing things for them rather than by what you see as sloppy or embarrassing displays of affection. Having said this, Virgo is an earth sign, so you may find it easier to show love sexually than by buying flowers or behaving in a romantic manner. In less intimate relationships, you make the most wonderful friend.

Occupations that suit you best are those that require attention to detail. Anything involving hygiene, food and diet, physical therapy or medicine is likely to attract you. Apart from these things as general interests, you can apply your energy towards helping those who cannot help themselves. In this regard, you have an innate knowledge of how to do the right thing without being told, and you also appreciate those who take your help and improve themselves as a result.

Because you tend to be rather shy and retiring, you don't seek the limelight, and you object strongly to those who attract attention when you know they have done nothing to deserve it. Even though you expect a lot of others, you set the same impossibly high standards for yourself and you are your own worst critic. You may be fussy and neurotic and make far too much noise over things like household cleanliness or what the neighbours think. Alternatively, you may be more relaxed than this, but your parents may have imposed exacting standards of this kind upon you.

It's likely that your parents were unloving or unaffectionate towards you, and your relationship with your mother may have been especially difficult. In many cases, the mother is neurotic, demanding, selfish and possibly even crazy. Often both parents have ridiculously high expectations and they criticize freely whenever they consider that you fall short. This is not necessary, as even the slightest adverse comment would make you hang your head in shame - whether it was deserved or not. When looking back honestly at your parents, ask yourself what their achievements amounted to, what they have done that was so wonderful. Also consider whether they were in reality failures in one way or another. Above all, you must learn to be kinder to yourself. You may reverse all this when dealing with your own children, being far too indulgent and ready to jump to their defence, thus allowing them levels of shoddy performance and behaviour that you don't allow yourself.

Relax by switching off from work or from other people and by taking a walk or reading a good book.

The Moon in Libra

The air element here makes it hard for you to integrate the emotional and restless nature of the moon, but the cardinal element helps you to get going when you need to do so. There are times when you would prefer not to have feelings and emotions at all but to work on a level of pure logic, but the moon intrudes on these and makes you feel hurt and pain whether you want to or not.

You are sociable, refined and sensitive to other people's opinions. It's important to you to be accepted by those with whom you have to deal in your daily life. Because you are anxious to maintain a middle of the road stance, you avoid antagonizing those who have different views and opinions from yours, although you can argue like mad when something is important to you. You don't care for bad language and vulgarity and you may prefer to walk away than to put up with this.

Because you are more sociable than romantic, you can easily be satisfied with intellectual companionship, where others would only be comfortable with physical contact. Although physical expression of your feelings is important to you, it cannot exist if you are unable to relate to someone in any other way. Alternatively, you may seek out sexual partners by the dozen but find their company boring or not up to the high standard that you believe yourself to deserve. You may be extremely loving, but there are some moon in Libra types who basically can only care about themselves, or perhaps their children. Even here, you may find it easier to show your love by showering them with expensive gifts.

Your refinement and sophistication mean that you prefer to associate with others who share similar tastes. Your charming manner, close attention to dress and kindness to others draws people to you. Occupations that suit your temperament are public relations, the law, any kind of artistic enterprise, the marketing side of sales, social organizations or elegant functions.

You may sometimes be unrealistic in your desire for a wealthy and comfortable lifestyle or cultured surroundings. Even when you under financial stress you are capable of acting the part of success so that you can keep up with the Jones's. You need a spacious and elegant home and you love to be surrounded by lovely things. One friend of mine spent a wet summer's day taking her children round a Georgian style stately home. She quite expected her six-year-old moon in Libra son to be bored by the outing, but he thoroughly enjoyed and appreciated the fine furniture, antiques and articles. Many people with this moon sign have two homes.

Some of you have problems when it comes to making decisions or choices, preferring to talk these over with colleagues and loved ones and to gain their approval before going ahead. Some of you have no difficulty in this area, but still try to please everybody or keep all the plates spinning on their poles, wearing yourself out in the meanwhile. The chances are that your parents were reasonable and your childhood happy, but they may not have enough money to give you all the things that you craved. As a parent yourself, you are far too indulgent with your own children, giving them more than they could ever need, but sometimes forgetting that what they really need is love.

Music, the arts and buying yourself attractive, indulgent or pretty things relaxes you.

The Moon in Scorpio

This water sign is very much in tune with the emotional and restless nature of the moon, but sometimes this can be too much of a good thing, which means that you can occasionally become over-whelmed by your own emotions. Your intense emotional nature means that your feelings are frequently expressed in extreme ways. Whilst you may sometimes fall in love to the point of infatuation, it's also easy for you to attract others and there are times when you positively enjoy teasing the people you entice. You may become defensive in your emotional dealings, and if someone doesn't give you the wholehearted adoration that you crave, you can become resentful and bitter. However, when you find yourself in a happy relationship with a partner who you can trust, you are loyal and faithful to the very end. When others are completely honest with you, there is nothing you won't do for them, and no sacrifice which you would not make. If you are undermined or cheated, you neither forgive nor forget.

You are possessive towards those who you love and you must guard against unreasonable jealousy. It's important for you to differentiate between those who are sincere in their relationships with you, and those who will never make any contribution to sustain a genuine relationship. You tend to be an opportunist, so you must also beware of expecting too much from people.

Avoid indulging in daydreams, because this may encourage you to lose contact with reality. If you overdo the importance of the present, you may fail to plan for the future. It's all too easy for you to get stuck in a rut, and then wonder why you aren't making any progress. You should try to identify your latent abilities, and utilize your talents by applying them efficiently to your goals. Once you put your mind to something you can see it through to the end though, so attainable goals are all important for you.

The water element of Scorpio encourages you to travel and you enjoy any opportunity to get away and discover new people and new places. You may also take inwards journeys, developing your spiritual and psychic side for your own benefit and for the benefit of others. You have a strong interest in psychology and also in healing, so you may take this up in the form of astrology, spiritual healing or by a career in the more normal routes of psychology or medicine. Your strong imagination could lead you to a career as a writer or artist, and expressing your creativity in this way will do a great deal for your confidence.

There is no reason to suppose that your childhood was nasty or tragic, but there was certainly something there that upset you. Somehow you feel as though you missed out on your proper share of love and attention. I know two people with this moon sign who grew up in households where a brother or sister could do no wrong and where they themselves were deemed to be second class citizens. Perhaps this is why you push others to the limit so that you can prove or disprove the reality of their love for you.

You may in your turn become a loving parent, but many moon in Scorpio people find it far easier to love animals than children - or for that matter, people. In fact, you find it relaxing and fun to play with pets and animals.

The Moon in Sagittarius

Sagittarius is a fire sign and the moon is a very different entity, but the restless and wandering nature of the sign actually suits the moon quite well. You may be a great traveller or you may simply prefer to allow your mind to explore and wander, but either way you are probably reasonably comfortable with your own inner nature.

You have a personality that loves freedom and that enables you to enjoy life to the full, and for the most part, you are optimistic and good-natured. Pessimists annoy you and whiners bore you. Part of the problem is that your nerves are more sensitive than others realize, and you have a tendency to sop up the emotions of others. This means that difficult or self-absorbed people can drag you down. You are constantly on the go, with an insatiable thirst for knowledge and curiosity about a wide range of subjects. Even though you are emotionally restless you want to form relationships with others, but they must accept that you will always be busy and involved with many projects. It's important for them to realize that you are quite happy with friends and lovers who share your acceptance of relationships that are not necessarily permanent or full time.

You get on best with professionals or people who have ambitions and who want to achieve success in life. You need to be with those who offer you variety and mental stimulation because idle chatter is of no real interest to you. You adapt easily to changes of environment, and although you may not travel to distant countries, you will certainly broaden your mind through books, the theatre and cultural interests. One downfall is that you might gain a reputation for intellectual arrogance or appear to be a know-all. Having said this, some of you do travel the world - probably as part of your working life. You certainly have friends among many different cultures and you are unlikely to be prejudiced in any way.

Some of you work in politics or religion, but many of you find your way into the fields of mind, body and spirit work as healers, astrologers, psychics and so on. You need to discover your own beliefs and philosophy, because you can't simply follow others in a sheep-like manner. You may also be quite eccentric both in your ideas and your lifestyle. You are attracted to show business, so there is a fair chance that at some point in your life you will appear on the stage or screen. Many of you work in broadcasting or have a regular spot on the radio or television at some point. Even if other areas of your horoscope make you shy and retiring you find it easy to hide behind your "stage" personality and temporarily become someone else.

Romantically, you are drawn to those who enjoy life to the full and whose goals include success and earning the money that assures a comfortable life. Those who live with you need to accept your inclination to be constantly on the move, because limiting your mobility would destroy your eager enthusiasm for life. You are an extremely loyal friend, but you may not make a success of love relationships, possibly because they become too cloying, demanding or intrusive. You are better at doing practical things for others than showing affection in a touchy-feely manner. You need time and space for yourself and sometimes a brief change of partner - especially when you feel in the mood for sexual experimentation.

There is a fair chance that your mother was unable to show you much affection and she may even have been demanding, eccentric, lazy or crazy. If you have children of your own, you will try to do your best for them, although your own somewhat eccentric lifestyle may make you a somewhat distant or detached parent.

It's said that Sagittarians love large animals and have a particular affinity with horses. The fact is that sun Sagittarians enjoy small animals and pets, it's the Sagittarian moon sign that's attuned to horses. A friend of mine once raised money for charity by giving readings in a "gypsy tent" at a horse fair and she found that all but one of her clients had the moon in this sign!

The Moon in Capricorn

The shy, reserved nature of the watery moon is reasonably happy in this earth sign, although it can make you a little too long-suffering for your own good.

You have a reserved, cautious personality which can lead to loneliness and you may not be able to derive satisfaction from relationships without feeling that you must pay some penalty. You badly want the ideal romantic relationship, but somehow you always seem to settle for second best. Your serious, ambitious nature is primarily concerned with achieving goals and objectives and this may be incompatible with concentrating your energies on domestic bliss. It's hard for you to project warmth and tenderness, so you attract those who are happy to accept what seems like a business relationship rather than romance. This is a shame because you really do want to express your true feelings with someone special. It may be equally hard for you to turn to others for love, affection and sympathy because you see yourself as a capable, efficient and independent person so such an admission appears to you to be an admission of failure. You may fall into the trap of thinking that it's enough to provide a partner with all their material needs and to neglect their emotional ones.

You may be pessimistic, far too materialistic and you may demand far too much in the way of standards of a partner. If you can get the idea into your head that a partner is neither there as a domestic machine or a wonderful money-earner, but simply someone to love, you will make things work. You must try to avoid fussy behaviour, or being prickly and taking offence over stupid things - especially when none is intended.

You are serious and hard working and more ambitious than you may appear to be on the surface. As you also tend to do things thoroughly - albeit at your own pace - you can stay in a job and slowly but surely climb the ladder of success. Where practical matters are concerned, you are sensible, able and careful, so whether you take on a task at work, start your own business or decide to rebuild your home, you will get there in the end.

The cause of your basic emotional insecurity may have been early parental conditioning, and your parents may have made you feel inadequate when you were young. Because of this, you are probably driven to accomplish professional or career objectives before getting too involved with close relationships or marriage. You really need to prove yourself and eliminate doubts about your own competence before you can relax. You also need to develop optimism and to allow your dry sense of humour a chance to see daylight.

Your background may have been loving, but your parents may have been extremely poor or there may have been too many children in the family for you to get your share of affection and attention. This may still make you shy, with a tendency to fade into the background, but it also makes you work hard for a lifestyle that's materially secure and for status within your community.

One odd thing, is that you may feel happier living on a hill than in a valley.

The Moon in Aquarius

This logical, analytical air sign of Aquarius is at odds with the watery, emotional nature of the moon. This makes it hard for you to cope with your own emotions at times and almost impossible for you to cope with those of others. The fixed nature of the sign makes you determined and obstinate.

You are extremely curious and eager to find out everything you can about people. Your feelings and your intellect go hand in hand, so that you try to understand those to whom you are emotionally linked. Your interests and your career will lead you into arenas where there are large numbers of people rather than small groups. You seek the truth in any situation, and sometimes this means that you feel let down by human frailties because the truth can be so painful. It's easy for you to be friendly with everyone, although intimacy might be hard for you, unless it satisfies your particular objective in life. Freedom is important for you, but you may not extend the same licence to those who you love.

Study, looking into different systems and discovering a philosophical truth can be an emotional experience that gives you considerable fulfilment. You are always very generous with the knowledge you acquire, sharing it with everyone who shows an interest. You have odd ideas about money and strange ways of dealing with it; on the one hand, you may not have much interest in money, preferring intellectual wealth rather than the material kind, while on the other hand, you may be extremely security conscious.

You have a huge amount of inner power and strength which can take you a long way in business or other worldly arenas. You don't like being second best or to allow anyone or anything to get the better of you, so whatever is on the outside of your personality, there is a deep level of ambition on the inside that leads on to success.

Yours is quite a tense moon sign, but you have a wonderful sense of rhythm and balance, so you can relax by dancing or by learning to play a musical instrument.

The Moon in Pisces

The watery, restless nature of Pisces fits the watery, restless nature of the moon very well, although the presence of such an emotional and intuitive planet in this sign can be too much of a good thing at times.

This is an extremely sensitive and vulnerable moon sign. You cling to those who give you emotional sustenance and even sometimes to those who don't, and you can hang on to relationships with parents and partners for far too long in the hopes that they will eventually give you the love and support that you need. You may eventually find it easier to love your children than other adults, and there is a fair chance that you will receive far more love back from them - or from pets and animals.

Sometimes the messages that you receive from parents and lovers are extremely confusing, as they may tell you that they love you while their behaviour demonstrates the fact that they do not. The old adage of "don't tell me, show me" is a good one to use here, as unless these people can demonstrate that they do actually care about you, you may be better off leaving them to their empty words. You must guard against a tendency to sacrifice yourself for selfish or destructive people or to base a relationship on fulfilling someone else's needs. At some point in your life, you experience a deep sense of loneliness. Sometimes this is due to childhood illness and time spent in hospital, otherwise it's the loneliness that can be felt due to being out of step with others.

Many people with this moon sign work in the mind, body and spirit fields as healers, astrologers, psychics and so on. It's easy for you to tap into the hurt and pain that others suffer and to try to heal this. You may even find yourself absorbing the feelings of others. One lunar Pisces friend knows that she can feel extremely angry, outraged, upset or depressed when someone else in the same room feels that way. This can make you confused about your own needs and feelings. Your heart is very soft, but despite this there is a core of strength and even of self-preservation that can save you from complete absorption or destruction.

Money is unlikely to be your god, because your values are spiritual than material. However, you can make money at times in your life, perhaps through a creative, musical, spiritual or artistic talent. Once you do this, you must guard against allowing other members of your family to spend it for you. You may be so self-sacrificial in this respect that you end up as a martyr. Oddly enough, some lunar Pisceans are extremely stingy and penny-pinching, either to themselves, to others or both.

While most lunar Pisceans are the soul of kindness, there are many who have an extremely spiteful and unpleasant side. This may only become apparent when you have to deal with them on something other than a superficial level, but it comes as a nasty shock. The root of this behaviour is a peculiar ego that is both insecure and lacking in confidence and at the same time superior and arrogant. The two fish of this sign make this type of personality to define or sometimes to live or work with.

Needless to say, your childhood was probably not that good. Your parents may have been too busy doing other things to attend to your emotional needs or they may have been critical and unnecessarily harsh. Your talents and your capacity for love are far greater than that of most other people, including your parents, but you need a partner and friends who support you.

Chapter Five:
Introduction to Lunar Houses

We are now entering realms of astrology that are really beyond the scope of this book. I will give you a simple system for working out your houses in a moment that will get you off the ground and allow you to look into your own particular lunar house. The fact is that this will really make much more sense when you have a proper chart and a list of planets and features or if you are already this far into astrology.

The astrological houses run in an anti-clockwise direction around the chart from the position of the Rising Sign. If you use the rule of thumb method that I now give you, and if you find that yourself dithering between two rising signs, try both and see what happens.

The Ascendant / Rising Sign

Your Sun sign depends upon your date of birth, but the Rising Sign depends upon the time of your birth. Once you have discovered this, it's easy to count off the astrological houses from this point.

- Look at the first illustration below. You will notice that it looks like a 24 hour clock showing the time of day in two-hour blocks.
- Place the astrological symbol that represents the sun (a circle with a dot in the middle) in the segment that corresponds to the time of your birth. If you were born during Daylight Saving or British Summer time, deduct one hour from your birth time. Our example shows someone who was born between 2 am. and 4 am.
- Place the name of your sign on the line at the end of the block of time that your Sun falls into, going in an anti-clockwise direction. Our example shows a person who was born between 2 am. and 4 am. under the sign of Pisces.
- Using the second illustration, write in the names of the zodiac signs in their correct order around the chart in an anti-clockwise direction.

- The sign that appears on the left hand side is your Rising Sign. The example shows a person with the sun in Pisces and with Aquarius rising.
- Now number the twelve houses, sign by sign, counting your Rising Sign as number one.

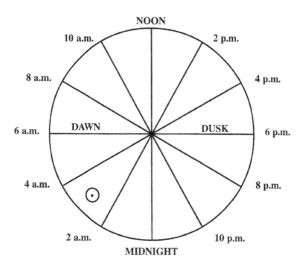

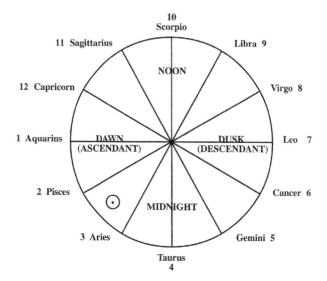

Chapter Six:
The Lunar Houses

The signs of the zodiac show how a planet behaves, but the houses are slightly different in that they show the areas of life that the planet focuses on. This can give a wonderful indication of what a person is really about and what they really want out of life, regardless of the impression that they give on the outside. The moon is so strongly influenced by the mother (or whoever took this role during a person's childhood), that it can show the beneficial effect of a good early role model or the harmful effect of a poor nurturer. The moon's house manifests itself in the adult's person's innermost and secret part of his personality. In fact, the sign, house, aspects by other planets, eclipses and much else will give a wonderful insight into what really goes on inside a person's head and heart.

The First House

When the moon is in this house, it was just about to rise at the time of your birth. This means that your mother is a very strong influence, for good or ill - or both. Your emotions are strong and you will need to find a way of expressing them. This means that you need a supportive partner who you can trust with your secrets and your innermost feelings. You also have a wonderfully creative imagination. You may be restless and keen on travel, but you also need a secure home, a family and a place where you can be yourself. You may have a strong ego or a need to fulfil your own requirements as well as those of others. It's possible that you will be interested in various aspects of food; perhaps what it contains, in addition to growing and cooking it. Another interest will be property, architecture and décor. Consider Prince Charles with his powerful mother-figure and his first house moon in the royal sign of Leo.

The Second House

This concerns personal values and priorities, so you must sort out what these mean to you before finding yourself spending too much time and energy providing for the needs of others. Money, goods, possessions, land and property are important to you and there is a chance that you could inherit these through the maternal side of your family. Art and beauty inspire you, so you may be keen on music and also those things that you can do with your hands such as gardening, cooking or do-it-yourself jobs. Due to your affinity with money, goods, services and property, you can work in fields that deal with these things. You may have been influenced by a particularly materialistic mother or family background, and it's possible that your family suffered from poverty and deprivation at some time in the past. This will drive you to ensure that you don't go without anything that you want during your lifetime.

The Third House

This relates to brothers and sisters, so these would have had a strong influence on you. This can make you especially close to your family or resentful and furious at them for past hurts and injustices. You are quite restless and you need variety in your life and in your job. Education and communicating comes easily to you, so this could lead to a career path. There is something childlike about you and you may never quite grow up. Your mother would have been keen for you to get a good basic education with plenty of intellectual stimulus, but your parents may have given you a limited view of the kind of education or career you should aim for. You may have problems with love relationships though, due to a need for personal freedom or detachment at one moment and a need for closeness the next. Alternatively, you may have experienced this kind of treatment when you were young. You are dexterous and you have a natural affinity for words and language.

The Fourth House

This is the moon's natural house, as it's associated with the sign of Cancer and the rulership of the moon in astrology. You may be very like your mother in nature or appearance and you probably were or still are very close to her. You appreciate her for what she did for you and you share her values. It's possible to work in a completely opposite direction, in that you have good reason to dislike your mother, but that's not the usual scenario. Home and family life are extremely important to you and you may even work from home or have a small business close by. Having said this, the restless nature of the moon can bring more than the usual number of house moves during the course of your life, and even emigration is possible. You can win and lose financially through property dealings.

The Fifth House

This house relates to creativity, so you have a strong urge to create. This may lead you into the arts or show-business, but it can equally mean the creation of a home, family or business. There is something childlike about you and you relate well to children and young people. You love games, sports, gambling a little and having fun. Your sex drive is strong and you may have numerous fun-filled love affairs. You may also look for a partner who can mother you. Your mother was either extremely loving and caring and a great organizer or a real drama-queen who was totally wrapped up in herself. Take care that you don't also fall in love with yourself.

The Sixth House

This house relates to duty. The old-fashioned meaning of this house was that of servants and masters, so it can relate to your attitude to those who you work for and those who you employ. In short, anything connected to work, chores and those things that need to be done is represented here. You may be a workaholic or you may make onerous demands upon others. Another factor is that of health, so you may be interested in health and healing or you may simply be neurotic about these things. Your mother may have supplied a hard-working role model and great encouragement or she might have been critical and destructive towards you, or perhaps she was just a fuss-pot.

The Seventh House

This is all about relationships and partnerships and it can be where you play parent/child games in your private life. You need to be loved, but you may have to search out the right person before you give your heart, because your own capacity for love is strong. There is a danger that you can look to your lover to validate you and to give you the approval that you need. This is fine if your partner does so, but not otherwise. This can be a good position for business and other kinds of partnership because you tune in well to the needs of others and you are curious about people. You will not wish spend too much time alone and you will always seek out others to work and play with. If your mother was a factor in your life, she would have either been a beneficial and benign one or an extremely argumentative and unpleasant one.

The Eighth House

This can be a difficult moon placement or an excellent one. On one hand, you stand to inherit property or the wherewithal to buy it, and your strong intuition can make you extremely successful in whatever work you take up. Your curiosity can be turned to good use in some kind of career. Difficulties might have come through sickness or death in your family or you may have felt jealous of a younger brother or sister who supplanted you in your mother's affections. On the other hand your mother may have been a positive influence on you, and she may have encouraged you to make the best of yourself. You have a talent for dealing with officials and with official matters, often on behalf of others and you can make money for others. You may be genuinely sexy or you may have a strong sexual aura or charisma about you that's attractive to others.

The Ninth House

Your restless nature may make it difficult for you to settle down in one place, or travel may be a feature of your life that is strongly connected to your relationships. Your mother may have come from a different culture or background to the one that you live in, so you may find it natural to live in two worlds. You have a strong need for personal freedom and a desire to work out your own philosophy. You tend to see things very clearly and there is a kind of honesty about you that makes it hard for you to wander from the straight and narrow or to be fooled by those who are dishonest. There is no great need for you to remain in close contact with your mother, and you may not wish to, but whenever you do, it's likely that you enter into a world that is not quite sane.

The Tenth House

Your mother may have had a strong personality and she certainly exerted a strong influence on you for good or ill. She may have been a hard-working and successful woman in her own right who inculcated you with good habits and the determination and thoroughness needed to get things done. On the other hand, she may have been stingy, dour, unloving, critical, hard, uncaring and emotionally destructive due perhaps to her own disappointments and frustrations in life. You badly need to make something of yourself, to earn self-respect and the respect of others. You are highly ambitious and you could be a very high-flyer indeed with a wonderful capacity for thoroughness, details, hard work and for hanging in there when times are bad. Your inner need is not necessarily for money or status, but simply to reach a position where you can approve of yourself.

The Eleventh House

Like the tenth house that precedes this one, your mother may have been a hard-working and successful role-model that leads you forward to success by encouraging you in all that you do. She would have encouraged you to get as good an education as possible and to be as happy as circumstances allowed. On the other hand, she might have been as critical and harmful as the tenth house mother. You have a somewhat detached attitude towards others and you may actually prefer friendships to close relationships that make you feel smothered. Your mind is alert and your ideas quite original and you know a lot about a great many odd and unusual subjects. You may be able to look as though you really care about people, but at heart you may be quite detached and not nearly as interested as you look.

The Twelfth House

The moon in this house can be similar to the reading for the moon in the first house, as it's close to the ascendant and it will have just risen. Your emotions are powerful and so is your imagination, and you may be prone to day-dreaming or have a rich inner fantasy life which can be turned to good use in creative or artistic pursuits. It's also possible that you have strong psychic or mediumistic tendencies. Your mother may have been a repressive, difficult, critical and hurtful influence, but she could equally have been a quiet and gentle person who encouraged you in all that you wanted to achieve. Health may be an issue in your life, either because you yourself suffer periods of sickness or due to other family members who need to be looked after.

Chapter Seven:
The Nodes of the Moon

Before the discovery of the workings of the solar system, it was assumed that the sun travelled around the earth. This pathway or trajectory of the sun is still used by astronomers as well as astrologers to plot the positions of the planets and constellations. This pathway is called the ecliptic, and the positions of the moon, planets and constellations of the zodiac are plotted longitudinally along its course. The moon and planets move slightly above the ecliptic at some times and slightly below it at others, and the moon's up and down motion in relation to the ecliptic is greater than that of the planets. The moon spends two weeks above the ecliptic and then crosses it on the way down, spending two weeks below the ecliptic before crossing it once more in an upwards direction. When the moon crosses the ecliptic in a northerly direction, this is called the north node. When the moon crosses the ecliptic in a southerly direction, this is the south node.

The movement of the nodes of the moon through the signs is somewhat different from that of the planets. For example, the nodes move backwards through the zodiac, but this backwards motion is interrupted every now and again by a short period of forward motion so the actual motion is forwards and backwards, with the whole sequence gradually moving back over a period of time. Some tables of figures give the average positions for the nodes, others include the forward shifts. The average tables are called mean and the more detailed ones are called true. The nodes spend about eighteen months in each sign before regressing into the next one. The figures below are for the north node only - the south node is easy to calculate, as it's always in the opposite sign. The first list shows the opposites.

NORTH NODE	SOUTH NODE
Aries	Libra
Taurus	Scorpio
Gemini	Sagittarius
Cancer	Capricorn
Leo	Aquarius
Virgo	Pisces
Libra	Aries
Scorpio	Taurus
Sagittarius	Gemini
Capricorn	Cancer
Aquarius	Leo
Pisces	Virgo

NORTH NODE: SIGN INGRESSES FROM 1930 - 2030		
North Node enters:	Aries	18 Jun 1930
North Node enters:	Pisces	6 Jan 1932
North Node enters:	Aquarius	25 Jul 1933
North Node enters:	Capricorn	12 Feb 1935
North Node enters:	Sagittarius	1 Sep 1936
North Node enters:	Scorpio	21 Mar 1938
North Node enters:	Libra	9 Oct 1939
North Node enters:	Virgo	27 Apr 1941
North Node enters:	Leo	15 Nov 1942
North Node enters:	Cancer	3 Jun 1944
North Node enters:	Gemini	22 Dec 1945
North Node enters:	Taurus	11 Jul 1947
North Node enters:	Aries	28 Jan 1949
North Node enters:	Pisces	17 Aug 1950
North Node enters:	Aquarius	6 Mar 1952
North Node enters:	Capricorn	23 Sep 1953
North Node enters:	Sagittarius	13 Apr 1955
North Node enters:	Scorpio	30 Oct 1956
North Node enters:	Libra	20 May 1958
North Node enters:	Virgo	7 Dec 1959

North Node enters:	Leo	26 Jun 1961
North Node enters:	Cancer	13 Jan 1963
North Node enters:	Gemini	2 Aug 1964
North Node enters:	Taurus	19 Feb 1966
North Node enters:	Aries	9 Sep 1967
North Node enters:	Pisces	29 Mar 1969
North Node enters:	Aquarius	16 Oct 1970
North Node enters:	Capricorn	5 May 1972
North Node enters:	Sagittarius	22 Nov 1973
North Node enters:	Scorpio	12 Jun 1975
North Node enters:	Libra	29 Dec 1976
North Node enters:	Virgo	19 Jul 1978
North Node enters:	Leo	5 Feb 1980
North Node enters:	Cancer	25 Aug 1981
North Node enters:	Gemini	14 Mar 1983
North Node enters:	Taurus	1 Oct 1984
North Node enters:	Aries	20 Apr 1986
North Node enters:	Pisces	8 Nov 1987
North Node enters:	Aquarius	28 May 1989

North Node enters:	Capricorn	15 Dec 1990
North Node enters:	Sagittarius	4 Jul 1992
North Node enters:	Scorpio	21 Jan 1994
North Node enters:	Libra	11 Aug 1995
North Node enters:	Virgo	27 Feb 1997
North Node enters:	Leo	17 Sep 1998
North Node enters:	Cancer	5 Apr 2000
North Node enters:	Gemini	24 Oct 2001
North Node enters:	Taurus	13 May 2003
North Node enters:	Aries	30 Nov 2004
North Node enters:	Pisces	19 Jun 2006
North Node enters:	Aquarius	7 Jan 2008
North Node enters:	Capricorn	27 Jul 2009
North Node enters:	Sagittarius	13 Feb 2011
North Node enters:	Scorpio	2 Sep 2012
North Node enters:	Libra	22 Mar 2014
North Node enters:	Virgo	10 Oct 2015
North Node enters:	Leo	28 Apr 2017
North Node enters:	Cancer	16 Nov 2018

North Node enters:	Gemini	4 Jun 2020
North Node enters:	Taurus	23 Dec 2021
North Node enters:	Aries	12 Jul 2023
North Node enters:	Pisces	29 Jan 2025
North Node enters:	Aquarius	18 Aug 2026
North Node enters:	Capricorn	7 Mar 2028
North Node enters:	Sagittarius	24 Sep 2029

Karma and Reincarnation

The theory of karma and reincarnation comes from the Hindu and Buddhist traditions. The idea is that we go through many lifetimes experiencing many different events. These lifetimes end when our souls gain enough knowledge and experience to take the first steps on the road to Nirvana. Many people believe that the troubles they experience in this life are a payment for sins committed in a previous life. This may be so, but the true belief is that we pass through all manner of experiences in order to understand the meaning of them and to learn what each one feels like.

Hindu astrologers call the nodes of the moon Rahu and Ketu, meaning the dragon's head and the dragon's tail. The idea is that the tail of the dragon is somewhat stuck in a previous life while the head is preparing for the next one. The astrological idea is that certain things that come easily to us do so because we learned them in a past life, while the current incarnation is filled with lessons that have been set or that we have chosen for the purpose of developing our souls. Therefore, the south node represents at least one past lifetime, while the north node represents the lessons and tasks for the one at hand. Another theory is that if we didn't learn the last lesson well enough, we are given it to do again. All this sounds as though heaven is made up of school teachers who hand out increasingly trying projects - but reincarnation is only one theory, and every religion or philosophy has its own.

You don't need to believe in reincarnation to understand the nodes of the moon. If you prefer, you can take it as read that you find it natural to do those things that the south node represents and difficult to achieve those things that the north node stands for.

Politics, Society, Timing and Geography

Leaving aside the ideas of karma, there is a completely opposite view of the nodes that some astrologers believe in and which also seems to work. This is that the north node represents those areas where the political or social atmosphere runs in your favour, while the south node does not.

Incidentally, it's usually easier to get something off the ground when a planet transits the north node than when it does the south node. This isn't to say that you should give up before even trying, but things do take longer and are more difficult when the south node is being activated.

The Parents and the Past

Another matter that seems to be attached to the nodes of the moon is the experience of being nurtured or mothered - by whoever did the mothering. Sometimes, the nodes tell us something about our own experiences as parents. Oddly enough, the signs that denote a difficult childhood when they are on the ascendant or as moon sign are not necessarily so difficult for the nodes. For instance, Gemini is not an easy sign to have on the ascendant, but a Gemini node is not bad at all. The attachment of the nodes of the moon to the ideas of the past and of one's parents are not so hard to understand because this is exactly what the moon itself is about, so its nodes are bound to have some connection. When planets make aspects to the nodes by transit or progression, events concerning parents (especially the mother) tend to come into focus.

Premises and Property

The premises and property that we occupy are also attached to the moon's nodes. This seems to apply both to home premises and also those that we may rent or buy for the purposes of running a farm or a business of our own or to let out for income. This is also not hard to understand, because the home, premises, small businesses, family, heritage and the past are all assigned to the moon, so these things are bound to have an influence on the nodes. The fact is that the connection between the nodes and premises is more likely to be seen (and felt) when planets make aspects to the nodes by transit or progression.

A Different View of the Nodes

Some time ago, I came across a book that went into oriental ideas about astrology. One of these ideas is that the south node is symbolic of success, whereas the north node is an area of failure, loss, obsession and addiction.

Astrological Factors

As we saw in the chapter on the zodiac, the nodes are always opposite to each other, which means that they share the same gender and quality, but not the same element. However, they do team up in a fire/air or air/fire combination, or an earth/water and water/earth combination. If you take the nodes of the moon as an indication of a change of karma from a previous lifetime, this means that there are still considerable similarities from one lifetime to another. However, whether this continues to work over a series of lifetimes is hard to say. If you are not keen on the theory of reincarnation, then you need only to take it as read that some of your talents, behaviour patterns and ingrained habits are fine just as they are, while others still need a bit of work.

Planetary nodes

All planets cross the ecliptic at some time or other, so they all have nodes. If you want to check out the nodes of the planets in your birth chart, you will need to use professional quality astrological software to find them. There doesn't seem to be information about these available online yet, so if you are interested in this form of astrology, why not set up a few charts for people who you know and check these nodes out for yourself?

Chapter Eight:
The Nodes through the Signs

North Node Aries - South Node Libra

The Aries lessons for this incarnation are self-determination, creativity and living in the real world. The Libran benefits that you bring with you are charm, the ability to relate to others and a longing for justice and for things to be just right.

The Libran influence from the south node can make you too ready to fulfil the needs of others and to ignore your own, which means that this incarnation represents a journey of discovery into who you are and what you need from life. Sure, there will be times when you want to suit others, but this should not always be the case or you will end up becoming a doormat or a martyr. You need some time and space for yourself. It's possible that you unwittingly compensate for being badly used by others in a past life by behaving in a manner that is too harsh or domineering. Once you understand that this is only a compensation for being too soft in a previous life, you will soon seek a middle way.

The Libran south node will give you an inherent talent for art or music and you may find it easy to create a lovely home and garden and to cook or make nice things. You could also seek out a creative career or one that takes you into the realm of beauty. However, laziness and self-indulgence may haunt you and prevent you from being as creative and efficient as you might be. Tap into the Aries energy to get down to things and to get them done. Relationships are another difficult area for you as you may find it easier to sit back and let others do things for you than to make an effort for yourself. Alternatively, you may allow others to lean on you and drain you. You need to develop the kind of leadership qualities that enable you to take a firm but fair line with others.

The Libran influence can make it hard for you to come to a decision about anything or to be absolutely realistic, and you may sink back into fantasy-land whenever you are under pressure. There are times when you are too detached from others and sometimes even from your own feelings. You may dislike being alone and you may have a habit of hoping that others will do your donkey work for you.

You may lack confidence at times, but you have a loving heart and you work hard to keep the romance alive in your relationship - and perhaps that's not a bad thing to carry over from a previous life.

Childhood and Nurture

You parents may have behaved in a confusing manner, switching between excessive discipline and laid-back over-indulgence. Perhaps one was a martinet while the other was saccharine sweet. There may have been some kind of fantasy played out in the family home, so it would have been difficult for you to sort out the truth from the mists of lies and half-truths. In later life, challenging transits to the nodes can bring all this back in some way so that you find yourself momentarily re-living the confusion and mixed messages of the past.

Property and Premises

The confusion and unreality of your childhood will have left you with an inner urge for some form of stability, so buying and hanging on to property will become important to you. You may even go so far as to keep your home in your own name rather than trusting a partner to keep up their share of the payments.

Social and Political

The society and times that you live through will be right for anything that is pioneering, new, innovative or inventive. You may do well by working in or for a large organization that works for the benefit of the public. Acting as an agent or work in a legal field may attract you, but it may not be successful for you this time around.

Karmic Problems

- You may dislike being alone.
- You may ask for advice and then resent it.

- You may spend too much on clothes and cosmetics.
- You may speak in a tiny voice, or a monotone.
- You may be a doormat.
- You may have an inferiority complex.
- You may feel sorry for those who don't have a partner, and wish to fix them up.

North Node Taurus - South Node Scorpio

The Taurus lessons urge you to make sensible priorities, to discover what is of real value and also to develop some genuine self-confidence and creativity. The benefits from the Scorpio south node are a capacity to think and feel deeply, but you also have intuition and a talent for investigation. A Scorpio south node may bring hidden resentments from a previous life. The chances are that you were undervalued or taken for granted last time around and others may have taken advantage of you. You may resent the success that others achieve and you may envy those who appear to have more in the way of material wealth, self-confidence or sexual success. You may have an unconscious urge to take the bull by the horns and ensure that you get the most and the best of what is available - whatever the cost to others. The sign of Taurus is concerned with material matters such as personal funds, property, land and possessions. This means that you must work for what you need and also conserve any land, money or goods that come your way. Good management is one of the lessons of this lifetime.

You should be an excellent businessperson, with a talent for buying, selling and negotiating on behalf of yourself or others. This time around, it will be your job to handle the budget, point out potential areas of loss and wastage to others and restore situations to stability. All this materialism is fine as long as you keep it in perspective and as long as you don't drive your loved ones mad due to a pursuit of wealth or a desire to save every spare penny. Alternatively, you may dislike manual or menial work and seek to start out at the top rather than to learn your trade from the bottom up. You may be contemptuous about money and possessions, but your life's lesson is to care about these things.

It's possible that you demand too much sexually. Sometimes a need for an unreasonable amount or type of sex goes back to feelings of being unloved and a need for your partner to prove love over and over again - or in strange and unusual ways. Power games related to sex or money are a very Scorpio scenario and one you need to draw away from. Try to find your own inner power source and to remind yourself that you are capable, worthy and loveable. Alternatively, others may seek to restrict your sexual activities or to prevent you from eating, drinking or enjoying yourself.

You may be fascinated by spiritualism and happy to fall into a trance at a moment's notice. You find it easy to investigate psychic phenomena and you may even carve out a career as a ghost-buster.

Childhood and Nurture
Confusion may arise over what is acceptable behaviour and what is not. One or both parents may have confused you by telling you to behave in one way one occasion and in a completely different way on the next. Or one parent may have been harsh while the other was indulgent and apt to encourage you to get away with murder while the other parent wasn't looking. This means that you have to work hard to find a middle way and to behave towards others in a fair and reasonable manner.

Property and Premises
This could be a lucky area of life for you and you may inherit property or the means of buying some. Alternatively, you work hard to make a nice home for yourself. Land is important to you, so living in town is less appealing than a rural area.

Social and Political

Anything that you can make or produce with your hands will do well for you, also tasks that involve the hands and the eyes, such as building, gardening, farming or decorating will succeed. Anything in the arts or the creation of beauty will go well. The things you do naturally but find hard to make a success of are psychology, social work, police work or a career in a hospital.

Karmic Problems

- You may have extreme views.
- You may sympathize with the underdog, envy success and have no real idea of who deserves what.
- You may waste energy in trying to change circumstances that are out of your control
- You may be too money minded or financially unstable.
- You may feel justified in continually criticizing others.
- You may only be prepared to give when you are sure that you can get something in return.

North Node Gemini - South Node Sagittarius

The lessons for this life are to communicate with others and to stay around long enough to make proper relationships. You may also need to gain knowledge and to pass it on to others. You are lucky in that Gemini and Sagittarius are quite similar in nature, so there shouldn't really be any harsh lessons to be learned. You benefit from a capacity for finding the right thing to believe in and being broadminded enough to look into ideas and beliefs that are different from those you were brought up with.

You may have been a little slow at school or out of step with the other children around you. The chances are that later on you turn to study or to interests that are a little different from the norm, perhaps becoming an expert in some specific field. You will always keep your mind busy by reading a great deal and keeping up to date on most things. You are happy to pass your knowledge on to others, and this could take you into a career in teaching or training. If you devote too much time to study or if you become too caught up in matters of religion, philosophy and belief you will bore others. The reality is that you will take an interest in something that takes you into the public eye and become far more successful than anyone could have foreseen during your childhood.

You need a mental as well as a physical connection in your relationships, so you must choose a partner who is happy to share your world of knowledge and ideas. Your need for freedom and variety may take you into a career that takes you away from home or that brings you into contact with many people. Many people with this combination find work in the travel trade. You may also have a natural talent for sports and a good sense of timing. You certainly have a loving heart and a good sense of humour.

One odd feature is that you may marry a close friend or a foreigner. You probably have better relationships with relatives than many other people do.

Childhood and Nurture
Your parents will have done all they can to give you educational opportunities and mental stimulation, and they would have been happy to talk about anything and everything. Your family may have come from a rigid religious background which you turn away from in later life. School may not have been a happy experience for you, as you seem to have been out of step with other children in some way.

Property and Premises
You will do what you can to keep a roof over your head, but you may move around quite a bit in your lifetime. It's likely that you will have property in more than one country or move from one country to another at some point in your life.

Social and Political

Communication is the name of the game for you in this lifetime. You will find it easy to study and to explore ideas, but heavy duty jobs such as being a lawyer, Church minister, a headmaster or school inspector won't work. Teaching, broadcasting, journalism, the travel trade or using the Internet for business will succeed. Also politics, show-business or anything else that communicates ideas and entertains the public would succeed.

Karmic Problems

- You may flirt, flatter and charm while revealing nothing of yourself.
- Others may hinder your freedom.
- You may find yourself among people who are not your intellectual equals.
- You will become overtired quite easily.

North Node Cancer - South Node Capricorn

The Cancerian lessons here are for you to enjoy family life and to take care of others without sacrificing yourself. The Capricorn past life experience gives you the benefit of being ambitious and the ability to strive for what you want in life.

You need a place to call your own and a family to surround yourself with. You may sometimes be aware of a feeling of inner loneliness that is left over from the Capricorn south node influence. There are times when you want to behave like a hermit, but the sense of warmth and protection given by being a member of a group is essential to your well being. Anyway, you are most likely to retreat into a shell when you feel hurt, insulted or upset.

The energies of the north node draw you into collective experiences, where you can be part of a group or your community. This can extend itself to wanting to feel part of a movement, a nation or the human race for a while. The cardinal nature of this sign gives you the drive needed to get things off the ground and it's hard for you to sit around doing nothing, especially when there are people around you who need your talent and energy. The easiest way for you to feel useful is to have a family around you.

Sometimes the south node intrudes in a particularly harsh manner, making you penny pinching and dour. Hard work is all very well, but there is more to life than this, so you should learn to let your hair down and to have some fun. The difficult south node can sap you of self-esteem or it can make you feel that you are only of value for what you do or for what you achieve. You need to develop a slightly more self-centred attitude at times so that you can concentrate on your own emotional needs and requirements, rather than to allow everyone around you to make use of you.

Sometimes this node position can give a longing for plenty of money, goods and possessions. If this is your requirement, then you may turn yourself into a workaholic while you try to achieve your ambition. The root of this need comes out of a past life where there was little security and too much time spent living on the edge of disaster. Relationships may be difficult for you at times if you are too ready to judge others and to criticize them, and you may be far too interested in control. Somewhere along the line you learned the lessons of discipline and self-discipline - and this is fine, but it can make you keen to impose your inner set of rules and regula-tions on others.

Childhood and Nurture
It's possible that some part of your childhood was spent in an insti-tution - perhaps a boarding school or even a home where rules and regulations were the order of the day. This can stunt emotions and stifle originality. If this was not so, your parents may have been crit-ical and judgmental, inclined to emphasize doing "the right thing" and living up to the expectations of others.

Property and Premises

You may be quite fortunate in this area of life as there is an indication that you may inherit property. If this is not the case, you can tap into a kind of inner antenna that tells you which kind of property to buy, renovate and sell at a profit. You are torn between the need to see your property as a home to live in and as an investment.

Social and Political

Anything related to property and security would succeed, therefore real-estate, insurance or even making and selling burglar alarms would do well. Large organizations, banking and highly structured business companies won't constitute a successful career path for you, but a small business of your own will, especially if your product reaches the general public. You can succeed in the food industry or anything that enhances family life. Banking and big business, accountancy and forestry will have been good career options for you last time around, but not this time.

Karmic Problems

- You may want a happy home life but feel uncomfortable when you get it.
- You may be shy, self-conscious and prudish.
- You may be stand-offish.
- You may have a tendency to test others in various ways.
- You may be nervous of new or different ideas.
- You may have a strong desire to be respected by others.
- You may suck up to those who you feel are your social superiors.
- You may be stubborn or a fuss-pot.

North Node Leo - South Node Aquarius

The lesson for this combination is to take life by the throat, overcome problems and to succeed. The Aquarian south node talks of social democracy, rule by committee, group activities and the greatest good. Laudable ideals indeed, but they are not what the

Leo north node is about. The fixed masculine fire sign of Leo is concerned with gaining a position in life and by starting and maintaining the kind of enterprises that give work to others. This can then form the backbone of a society that has something to offer and that can afford to supply the means of support to those who need it. This powerful north node sign offers plenty of challenges for you to face during your lifetime, but even if you don't turn into a captain of industry, you can do something just as creative in a much smaller way.

Your home and family are desperately important to you and if there is even a hint that these are likely to be stripped from you, you will fight to the bitter end to prevent this from happening. You need to possess your own realm, whether this be the home and family, a business or part of someone else's business that is your own. There is always a degree of untapped creativity here – sometimes caused by genuine shyness, and it takes time for you to make use of this and to become what you need to be.

It's probably just as well that you enjoy a certain amount of drama and are happy to take centre stage, because fate will put you there and hope that you can cope with whatever may happen. The world will provoke confrontations and the node in this position will encourage you to grow and develop as an individual.

Sometimes traumatic experiences will cause you to retreat from the world while an ingrained snobbishness can hinder you when you wish to form relationships. Sometimes, it takes time before you can learn to love and trust a partner. It's quite possible that your first serious relationships are with those who don't understand you and cannot respect what you try to achieve. With luck, you find the right person later in life and then settle down to a far happier life within the home and the family.

The chances are that you are an extremely loving and affectionate person, but you must guard against throwing your loving heart away on those who don't appreciate you or who make use of you. It will take time for you to respect yourself enough to choose a lover who wants you as much as you want them and who loves you as much as you love them.

Childhood and Nurture

You may not have experienced a great deal of nurturing when you were young and there is a fair chance that you found yourself in situations that were extremely difficult to cope with. Sometimes this is a result of death of a parent or one who abandons the family, either by walking out or by taking little part in family life. Perhaps somewhere along the line there were people who didn't treat you with respect, and it may even have been difficult for you to respect yourself or to develop self-esteem. You do achieve this in time, but only after travelling a long way down a hard road.P

Property and Premises

You may have big ideas about the kind of home you would like, but you will probably have to make do with something less grand than you prefer. However, you are a practical person, and as long as you and your family can be happy and comfortable, this won't matter all that much. There is no specific luck attached to buying and selling property for you, so don't speculate in this area of life - rather find something you like and live in it.

Social and Political

It doesn't really matter what kind of society you live in, you will find a way to be creative and to make a living. You can tap into the things that people need and want and then make or sell the goods that work for you or provide a service that people require. Altruistic jobs such as social or charitable work are alright in their place but not as the main occupation, as this will not fulfil your needs this time around.

Karmic Problems

* You may spend too much time sorting out other people's problems.
* You may wish to be in the spotlight but find yourself being sidelined.
* You may talk about equality but in reality wish to be the boss.
* You may use your intellect at the expense of your emotions.
* You can be spiteful, especially when hurt.

North Node Virgo - South Node Pisces

These two node signs share certain similarities, in that they both talk about sacrifice and the needs of others rather than the needs of yourself, so there is a kind of double-whammy karmic lesson to be found at work here. You will need to take a realistic attitude to the needs of others, because everywhere you look, there are people who need help, either in a charitable sense or within your own family. While it's wonderful to give time, consideration and energy to others, you will need to bear in mind that if you don't look after yourself, there will soon be nothing left for you to give.

Health problems can indicate an imbalance and these can be physical or mental, real or imagined. You fear losing your independence and becoming dependent upon others as a result of sickness. In extreme cases, you could become a spiritual hypochondriac, drifting from therapist to guru, searching for someone to sort out your problems.

Being a flexible, mutable node sign, you seek to fit in with others, but you will also need to guard against losing your identity. You may have an excessive tolerance for unacceptable behaviour from others and faulty reasoning that allows you to gloss over their flaws. The danger is that you will spend time and energy helping others while your own life disintegrates. You would benefit by tapping into the fantasy and the imaginative creativity of the south node in order to work on artistic, musical or literary projects. If you can channel your energies into work that you enjoy and if you can find a partner who respects you and validates you, your health will improve dramatically. Guard against picking chaotic partners who need to be rescued from the messes that they themselves cause, because they will hurt you and then walk away whistling a happy tune while seeking out another muggins.

The key for your node combination is discriminating service to humanity - with discriminating being the operative word. You have an inner desire to express divine love to humanity, but you probably need to learn to love yourself first.

Childhood and Nurture
Your childhood may have been fairly ordinary or it may have been difficult, but either way, the chances are that you left home early in order to set up a family of your own. If this worked out for you, then your subsequent life would have been quite reasonable, but if not, you would have had a long period of struggle before finding peace and happiness.

Property and Premises
This is likely to be a fairly lucky matter for you, because there is a chance that you will inherit property or the wherewithal for a good home. Even if this is not the case, you will find a way of buying a home and eventually of paying a mortgage.

Society and Politics
You are inclined to fit in to whatever society you find yourself living in rather than behaving in a rebellious or outrageous way. You can always find a way of making a living and your modest, quiet and helpful behaviour will endear you to your neighbours. Work in the fields of health and healing can succeed for you, but although you may be quite psychic, this is not likely to be a fortunate career option.

Karmic Problems

- You may indulge in excessive daydreaming.
- You could give an appearance of being too sweet for words.
- You may have a strange or fluctuating attitude towards work.
- You may not exercise enough control over your emotions.
- You may be moody, to the point where even you don't know what's eating you.
- You may attract disorganized people and chaotic conditions.

North Node Libra - South Node Aries

Aries is the most self-centred of all the signs, while Libra focuses on partnerships, relationships and joint-ventures. The lesson is one of cooperation, sharing and all that goes with being part of a partnership with someone else in your personal life or your business world.

Diplomacy figures strongly in the Libran approach to the world, always in the interests of harmony. You can make compromises in order to be able to come out on top whilst at the same time looking good about it. Taken to extremes, you are capable of saying one thing to one person and saying exactly the opposite to another or of agreeing with the last person you talk to. On the other hand, both these signs can be argumentative, so you might ruin your relationships with others by disputing everything or by picking fights. If you enjoy argument and confrontation, take a job that requires this kind of approach, because if you insist on making yourself difficult to live with or be with, you will soon be alone for good.

It's important to strike a balance between what you need and what you need to give to others in order to live in harmony with the world. This requires a measure of self discipline and perhaps holding emotions in check so as not to be thrown off balance or be carried away. The north node here is a restraining factor that prevents you from behaving in a manner that is either too demanding or too eccentric. The pull from the south node will make you competitive, keen to prove yourself and you may sometimes be uncomfortable in social situations. You will have to find a balance between necessary assertiveness rather than aggression on the one hand, while not becoming a doormat on the other hand. There is a difference between caring for people and controlling them and you must allow others the chance to make their own decisions.

You may be inclined to complain that you have been forced into something that ruined your life. This could be something that you feel that your parents did, a wrong career choice, having a child that you were not ready for it, or an unhappy early marriage. You may even use this as justification for reluctance to co-operate with others or a reluctance to share. Aries wants what it wants quickly; Libra likes to take its time, so you will have to find a way of accepting that an easy jogging speed is best.

You can ally your competitive streak to your artistic talent and eye for beauty in such fields as, fashion, make-up artist, architect or interior decorator. Alternatively, your Arian courage can take you into a career that demands strong nerves.

Childhood and Nurture
Your parents probably did their best for you, but there seems to have been a history of rows and arguments. It may be that your parents fought with each other or that one or both of them found you irritating, and it's likely that they tried to talk or ague you into doing things their way. Yours is not really an obstinate nature, but it's a cardinal one which implies that you have to find your own way of going about things and that you don't much like dancing to someone else's tune. It's possible that one of your parents was a disciplinarian, but it's equally possible that one of your parents was missing altogether. If so, those who took care of you may have laid down rules and regulations that you couldn't live with or perhaps their view of life was completely different from yours.

Property and Premises
There is no special karmic benefit or karmic challenge in this area of life. You enjoy having a nice home and there is no reason why you shouldn't be able to have one.

Society and Politics
You have an instinct for what will work, and the charm to bring it off, so whether you want to become the manager of a pop group or run a large organization, you will find it easy to tap into the collective consciousness of the society that you live in.

Karmic Problems
- You may suffer ailments that result from frustration or rage that's driven inwards.
- You may switch between impulsiveness and hyperactivity to somnolence.
- You may display a cold attitude towards others' accomplishments.
- You may be argumentative.
- You may boast about your accomplishments and be a poor loser.
- You may always want to be the boss.
- You may find it hard to finish what you start because you bore easily or have a short attention span.

North Node Scorpio - South Node Taurus
The karmic lesson here is all about sharing. Your south node impels you to look after your own finances and possessions and to make sure that you are safe and secure in every way. This is fine as far as it goes, but if you are to live with and work among others a bit of give and take will work wonders. Well, the old wedding vows used to say "for better or for worse" and that's what a real relationship is all about - even in a way, those that we have with our friends and colleagues. There are times in every relationship when you have to put your own needs and feelings on hold for a while and support

someone else's. This doesn't mean living the life of a martyr or becoming so ground down that deep, unexpressed resentments make a mockery of your life.

You may play power games with others or you may be on the receiving end of these. The kind of relationship where one person isn't allowed to smoke, drink, keep a pet or eat onions could be involved here, or there might be excessive criticism. If you are on the receiving end of this, you will have to learn tactics such as changing the subject, not rising to the bait and of answering unpleasant remarks in a mild tone of voice. If you dish out hurt to others, you will have fun for a while until they leave you in preference for someone kinder.

You may find that you have to act on behalf of others or take care of other people's money, goods or property at some point in your life. This requires standing back from the situation while you work out what is required of you and the duty or obligation that you have to the situation. It's possible that you can make a career in something like banking, accountancy or legal work where you look after the money or interests of others in a professional capacity. You may be drawn to serious jobs such as police or medical work where you look after the needs and requirements of the public. Even something like a job where you shepherd children across a busy road outside their school is a possibility, as here you are looking after the off-spring of others. Another possibility is that you will need to keep other people's secrets at some point in your life. This may be due a personal situation, but it could be part and parcel of your job. If you work in some kind of public sector job, such as the Social Services, the Police, education, psychology or health, you will definitely be handling confidential matters.

Taurus is quite a self-indulgent sign, but Scorpio is a far more hard-working one, so you may find yourself lumbered with more chores and duties than you can cope with at times. Try to organize your life and your day so that you can get everything done. You are bound to find yourself overwhelmed at times or living in some kind of muddle where you can't find a home for all the things you have, so you will have to work out space-saving or time-saving ploys in order to put your life into shape.

You have a talent for investigation which you might put to good advantage in a career, either in something like the Police, forensic medicine, medical research, surgery or even psychical research.

Childhood and Nurture
This combination doesn't indicate a really difficult childhood, but somewhere along the line you will have learned how to keep things to yourself. This may have been a form of self-protection, or perhaps your family knew things or did things that they didn't want others to know about. You may have felt resentful of a cleverer, luckier or more popular brother or sister and it will not have been acceptable for you to complain about this. This can in turn make you critical, jealous and spiteful to others in later life.

Property and Premises
You may be quite lucky in this area of life, either due to inheriting money or because you work your way up in a job and put your money into a nice house. If you avoid divorces and so forth, you will be able to pay off a mortgage and have total security in later life.

Society and Politics
There is nothing to suppose that your lifestyle or thinking processes are at odds with the society that you find yourself in. In fact, the chances are that there will be plenty of opportunity to find work in some sphere where you act on behalf of others or take responsibility for the possessions of others. You should be in tune with the time and place that you find yourself living in.

Karmic Problems
- You may refuse to believe what others tell you, even when you know they are not liars.
- You can be insensitive and forget that others also have rights.
- You may be shy or introverted.
- You may view everything in terms of money
- You may be self-indulgent.
- You may cover up cowardice by bragging or bullying others.

Sagittarius North Node - Gemini South Node

These two nodes have a fair bit in common, so if you believe in reincarnation, there is a feeling that this lifetime is a case of finishing up unfinished business rather than tackling completely new issues. Both these signs involve communication and education, so you will need to gain as much education as you can this time round. You may do well at school while you are young, or go into some form of education later in life, and it's quite possible that you will find work as a teacher yourself. This doesn't necessarily mean that what you study (or teach) will be a mainstream subject. It might be, but it's equally possible that you will take an interest in philosophy, astrology, spiritual matters and what could be termed new age subjects.

Gemini is all about local matters while Sagittarius needs to travel far and wide, so you may travel as part of your work or in order to find your destiny. It's possible that you will leave the country that you grew up in and choose somewhere else to live. You are broadminded and not at all prejudiced towards people of different races, religions or cultures; indeed you find them all fascinating. This could take you from a conventional home and lifestyle, to the Zambezi River, the jungles of Borneo or anywhere else that captures your imagination.

The chances are that you will question the religious beliefs or the accepted beliefs of your parents, family, school or anybody else in your childhood surroundings. This will send you on an inner journey and an outer quest to find something that appeals to your sense of what is right and what works for you as a belief system. Your sense of justice and need for honesty could take you into a career in the law or even something to do with the legal system of a country.

Issues relating to personal freedom come to the fore at some point in your life. On the one hand, you need to be able to come and go as you like. This may even be essential as your job could be the kind that takes you from one place to another, but you will want your partner to be home-based.

Your previous incarnation may have inclined you towards practicality, but this one will take you down a mystical path, in which spiritual ideas and such things as healing, channelling and learning to divine for water in the middle of the desert might take precedence. Banal, stupid or prejudiced ideas won't wash with you now, because you filter these through your own mind and heart to see what makes real sense to you.

Childhood and Nurture
You may have been an odd-man-out during your childhood, either because you didn't quite fit into your family circle or because you were out of step with the beliefs and ideas to those of the other children in your neighbourhood - even possibly your teachers. It's possible that your parents were immigrants who themselves didn't really understand or become a true part of the community you grew up in.

Property and Premises
You are not terribly interested in property matters, but you would like a home base to come back to. There is no reason to suppose you will lose out in this area of life.

Social and Political
Whatever the prevailing views are, yours are different. You will have to search for friends, colleagues and even perhaps a new place to live where you can explore ideas and formulate a belief system or a way of working that suits you, as well as the society that you live in.

Karmic Problems
- You may want more from life than it's possible for you to have.
- You may mess up your schooling and have to do it all again later.
- You may want others to provide the good things of life for you.
- You may have to learn a new language.
- You may have to retrain for a new career later in life.
- You may do things too quickly and not thoroughly enough.
- You may not be on the wavelength of others.

Capricorn North Node - Cancer South Node
The karmic lesson here is to depend less on others and more on yourself. If you believe in reincarnation, the last time round you would have been looked after and nurtured by your family, but this time you are out on a limb - for a while at least.

You may wish to lean on others, but the Capricorn north node is all about independence, so somewhere along the line you will learn that the only person you can be absolutely sure of is yourself. This may make you too cautious and slow to trust others, but perhaps this is how you need to be. In your personal life, you will have to seek out those people who you can trust and relax with, so that you can eventually enjoy a loving and kindly personal relationship.

Once you have achieved this, you can create a family of your own to love and look after.

You have a natural affinity for big business which means that you may work in one of those fields where you make authoritative decisions and have meetings with those who work with or under you. You may start out in a fairly lowly position and work your way slowly up the ladder of success either by staying put or by moving carefully from one job to another when the time is right. You will have to study accounts and figures in order to make the kind of decisions that affect other people or that make money for your firm or organization. The only drawback to all this is that you may become so wrapped up with your work that you neglect your family and ultimately lose them. You need to strike a balance here.

Your parents will be important to you, and it's quite possible that you will take on some form of responsibility for their welfare, either by providing for them in some way or by keeping in close contact with them. Both older and younger generations will figure heavily in your life. Cancer is all about motherhood and family life while Capricorn is all about fatherhood and bringing home the bacon. The karmic lesson seems to be to look after yourself and others this time round rather than to be looked after.

Childhood and Nurture

Your parents will have done all they can for you, but the circumstances of your childhood might have been difficult. It's possible that one of your parents died or walked out of the family, leaving the other one to struggle along alone. Another possibility is that you were part of a large family. Perhaps poverty was a problem. You may have been very quiet and shy or it may have been circumstances that prevented you from expressing yourself or putting your needs first.

Property and premises

Inheritance is possible for you, but it would only come later in your own life. In the meantime, you work hard to build and maintain a home for yourself and your family. You may also have to subsidize or look after the home and surroundings of other family members at some point in your life.

Social and political

You will fit easily into whatever culture you find yourself in and you will take the opportunities that are offered to you in order to make your way in life.

Karmic problems

- You may be moody and depressed at times.
- You may lack self-confidence or feel that you are unsuccessful.
- You could be stingy or penny-pinching.
- You so attached to your mother or family that you cannot break away from them.
- You may sacrifice opportunities for the sake of loved ones.
- You may make your career the centre of your life and ignore your loved ones.

Aquarius North Node - Leo South Node

Leo is self-centred while Aquarius is all about the greater good for the largest number of people. A Leo incarnation is all about self-development, creativity and learning to love and respect oneself, but an Aquarian one is quite different as it refers to a love of humanity or perhaps even of the planet itself. You may find yourself drawn into a cause which is for the benefit of mankind, and even if some hangover from your Leo south node thrusts you into the limelight, it will be on behalf of a cause than for your own glory, benefit or financial gain. Both of these nodes are quite proud and even arrogant, and you may display intellectual arrogance this time round and you could even feel superior to others. On the other hand, you may be so keen to do the best you can for your community or perhaps for the firm that you work for that your own achievements get overlooked. Others who are quicker on the uptake than you may march in and grab the glory that should rightfully be yours. You may be more inclined towards the realms of ideas and ethics than making money, making your mark on the world or indeed doing anything practical.

Personal relationships are fairly easy with this combination, although these will be happier later in life than they are at the start. The chances are that you will go through a couple of deep relationships that don't work out before you find the kind of person who will put you first - or at least be prepared to make a true partnership. Friendships will always be important to you, and some kind of inner need for freedom and detachment may make you prefer this kind of detached relationship to the closer type. One good thing that you transfer over from the Leo node is the ability to have fun.

You have a streak of originality that you can put to good effect, either by inventing new products or by coming up with new ways of handling old problems. Both Leo and Aquarius like new technology, so you may be drawn to gadgets and electronics as part of your career. Teaching and training also attract you because you enjoy developing the talents and intelligence of others.

Childhood and nurture
There is nothing to suppose that your childhood was difficult, but you developed a certain amount of independence and perhaps stood on your own feet from an early age. Your parents may not have been able to take complete care of you, so it's possible that you spent more time at school or among friends than other children did. You may eventually turn your back on your family.

Property and Premises
You may not be all that interested in property for the sake of its value or as an investment, so you may prefer the freedom of renting rather than buying. Alternatively, you may have a property of your own in one place but rent another for your work.

Social and Political
You may buck the prevailing trends in your area of birth or life and therefore, need to move to another area or even another country in order to find opportunities and outlets for your talent.

Karmic Problems
- You may not wish to grow up and accept responsibility.
- You may be arrogant or snobbish.
- You may act in a childish way.
- You may want to take centre stage when it's not appropriate.
- You may want children but find it inconvenient to have them.
- You may be more successful in groups or with friends than in family life.

Pisces North Node - Virgo South Node
There are similarities between these two nodes in that both are dedicated to service to others. This may lead you to sacrifice some part of yourself for the sake of your family, friends or your community. The Virgo node relates to work and duty while the Pisces node has more to do with bringing spiritual sustenance to those who need it. This may take you into the world of caring for those who can't care for themselves, so you might find work caring for children, the elderly or the less advantaged members of the community. Pisces relates to places of seclusion and protection, so you may work in a hospital, an old people's home, a prison or a hospital for people with mental problems. The chances are that you will gain great satisfaction by making others happy and comfortable - as long as they appreciate what you do for them.

This is all very well, but sacrifices should be kept to a minimum or you will wear yourself out on behalf of others and leave nothing for yourself. If you have the kind of family who are happy to leave you to take responsibility for caring for the children or elderly of the family, this can also exhaust you and leave you with no life of your own. You must learn to put your foot down and to have some respect for yourself and for your own needs. Also, if your relationship is truly awful, you must analyse this to see if you are staying in it simply because you fear the idea of living alone.

Another key idea with this node combination is that of escape. You certainly need to escape from the demands of others from time to time, but you also need to consider the route that you take for the purpose. Building in holidays and times for doing the kind of creative things you enjoy is the answer, but some people with this combination escape into a world of drink and drugs. There is also a strong need to delve into the spiritual and mystical sides of life and to understand those things that are hidden, unusual or beyond the scope of the five senses. You may be happier delving into the meaning of life or spiritual matters than coping with practicalities - or vice versa.

This incarnation might take you travelling or it may bring you into contact with people of a different race, colour, religion or culture than your own. You would thoroughly enjoy learning from people from diverse backgrounds. Another possibility is that you take up an interest in health and healing, either because you have some sickness yourself that you need to learn to cure or on behalf of others. You may simply be somewhat neurotic about health or perhaps a stressful situation makes you sick at some point in your life.

Childhood and Nurture
There may have been an element of loneliness in your childhood. There are many possible reasons for this, such as being sick and in

hospital for a while or simply because the circumstances of your home and life at the time meant that you were isolated from others in some way. Any such isolation would have been beneficial, because it would have allowed you space in which to develop your imaginative and creative talents and possibly also your psychic powers.

Property and Premises
You may move around too much to acquire property or premises, but if other factors on your chart encourage you to have a fairly stable lifestyle, you will be able to settle down and buy something for yourself. You don't have a great business head on your shoulders, so you must take advice and take great care when buying or selling property.

Social and Political
You are so adaptable that you can fit in to most prevailing social and cultural conditions and make something of yourself within whatever surroundings you happen to be. If you don't like the place or the culture of your childhood, you will move out and explore the wider world until you find one that suits you.

Karmic Problems
- Your mind may be on overdrive for much of the time.
- You may be a perfectionist or a fuss-pot.
- You may talk too much or too loudly.
- You may find it easier to love animals than people.
- You may sacrifice yourself too readily and become worn out.
- You may attract chaotic people or you may have a chaotic lifestyle.
- You may escape into dreamland, fantasy-land and never get anything done.
- You may be too critical of others, and you may try to damage the reputation of those who you feel threatened by.

Chapter Nine:
The Nodes through the Houses

While the signs of the zodiac show what a node is like, the houses show where and how it operates. For example, an Aries north node belongs to an extrovert who doesn't let the grass grow under his feet but, if this is in the fourth house, the person's energies will be expended on his home and family and perhaps a small business rather than by climbing Everest or taking the world by storm.

North Node in the First house

This is an extrovert house, so there is a need for recognition; and if the node is close to the ascendant, you may achieve fame and fortune. Sport may be an excellent outlet for your energies and this will give you a chance to shine. You need to understand your own needs and to chart your own course in life. The seventh house south node may make you wish to lean on others at times or to allow them to put a value on you that suits them or to allow others to express opinions about your performance and thus to sap your self-confidence. You need to stand on your own feet and to make decisions and choices based on what you need rather than to lean on others or to be pushed around by them. It's best to avoid taking the line of least resistance or to sacrifice too much of yourself in order to fulfil the needs of others. You may over-compensate by becoming self-centred, so you must try to find a middle way.

You have great energy and enthusiasm and you can be a pioneer in any field that you take up. Being courageous, you tend to look for jobs that are difficult or even dangerous and you don't flinch from hard work. You are idealistic, so you are bound to want to do something that betters the lot of those who need help.

North Node in the Second house

This house is all about personal values and personal possessions. You will work hard to make a nice home for yourself and for your family and to own those things that make life comfortable. You may be a little self-indulgent and sometimes self-centred, but you mean well. A need to create beauty and to make things look nice might lead you into jobs that involve improving the surroundings or the appearance of others. If you can turn your hand to such things as singing, playing music, gardening, painting or creating beauty, you can be happy.

The eighth house south node means that you have deep emotions that are hidden under the surface and which can emerge at times to make you feel angry, resentful or depressed. When this occurs, try to do something practical rather than wallowing in unnecessary emotion or moodiness. You may fear poverty or being drained by others, or you may be too ready to rely upon them to keep you. Try to maintain a reasonable attitude. Don't allow others to drag you down; try to keep your spirits up.

The only other drawback to this node position is that you may be so security conscious or money-minded that you become tight-fisted or too materialistic. You may also envy those who are better off than yourself.

North Node in the Third house

The third house is primarily involved with communicating, and this may become part of a job or a way of life. Alternatively, you may become actively involved in local matters and the politics and events that happen in your neighbourhood. Your phone and email will be busy, and you will have plenty of people popping in to see you - that is when you are not out and about running errands all over the place. Education may be a feature of your life, either because you enjoy studying or because you find a career in teaching. Another common scenario for this node is to write for a living or to work in a field of communication - or oddly enough, of finance.

The ninth house south node may make you a little dogmatic at times and sure that you have the only opinion or belief that's worth having. Try to ease up and relax a little. Sometimes you will have to prevent your tongue from running away from you and letting the cat out of the bag, so take care with things like gossip or rumour-mongering.

You are happiest when surrounded by people who like you and who validate you, but you can attract jealousy, spite and users, so take care to trust only those who are worthy of this. Brothers, sisters and neighbours will figure in your life.

North Node in the Fourth house

Home and family life will be important to you and you will do all you can to keep the home fires burning and the family together. Hopefully, you allow your children to grow up and become independent, knowing that they will always keep in touch with you and be happy to visit you or to have you visit their homes. If you try to cling to them or to manipulate and dominate them, you will end up losing them.

There is a strong chance that you will work from home or run a small business of your own. You can fit in at a large organization, as long as you have your own job to do there and your own place in which you work. If people insist on moving your stuff around or borrowing things without telling you - and without putting them back, you can become extremely aggravated. You are most comfortable in a workplace that has a family atmosphere, but if competitiveness and unpleasantness are the order of the day you will eventually leave and set up something of your own. The tenth house south node makes you more ambitious than people realize.

North Node in the Fifth house

Creativity is the key to this house as you simply must have a creative outlet in order to be happy. You may be drawn to the world of show-business, fashion or some other form of glamour. It's no good expecting you to be a backroom boy, because you must shine in your own right, so you could be drawn to make a career in sports or in the world of entertainment where you can please and excite others.

You need to feel close to others to obtain sense of warmth. Perhaps this need for closeness draws you to love affairs that have a strong fun element and a transient feeling of connection. The south node is detached and prefers to remain at a distance from others so your own needs can be confusing to you. You will certainly wish to take risks at times, either in connection with the search for love or perhaps by speculating and gambling. Your other love is for children, so you may become devoted to your family, and you may choose to do something that involves children, such as teaching or running a boy or Girl Scout operation. The eleventh house south node makes you want to do something for the community at large, but this can only come along after you have established yourself and gained a reputation.

North Node in the Sixth house

The message here is to work for what you want and not to escape from the realities of life. However, you must also guard against allowing others to leave everything to you while they live in a fantasy land where everything gets done by magic. You need to ensure that you are given the recognition and rewards for what you do and that your good nature is not taken for granted.

The twelfth house south node gives you an inherent creativity which you can use for art, writing or perhaps song-writing, while the north node allows you to make order out of chaos. This is a powerful combination that can lead to great success, especially if there is a planet contacting one of the nodes. In business, you understand the principles of supply and demand and also those of satisfying the demands of customers. You make an excellent employee as long as you don't do everybody else's job as well as your own, and you make an excellent employer, as long as you ensure that others pull their weight.

You may be interested in health and healing, and the twelfth house south node may give you an inherited talent for medicine, nursing, psychic work or spiritual healing.

North Node in the Seventh house

The north node in the Seventh house is great if you need to make contacts and deal with others. Issues of commitment, marriage or business agreements are important to you, but you must ensure that these benefit both parties. However, it may only be you who is prepared to make these commitments, so ensure that you are not taken for a ride by others.

The presence of the south node in the first house means that you may fall into the trap of believing that everything should revolve around you. This can be a good thing while you are single or if you work alone, but this can be hard on those who have to work or live with you. On the other hand, you may be too quick to give in to others and unable to stand up for your own rights. A middle way is best here if you can achieve it. You would love others to validate you, but they may be reluctant to do so. You enjoy encouraging others to make a success of themselves, and sometimes this leads to them taking advantage of your good nature. You are refined and pleasant and you prefer to live in peace, but arguments and disputes may follow you around.

North Node in the Eighth house

The eighth house is not an easy one for a beginner to understand. Astrology books tell us that it rules birth, death, sex, shared resources, business partnerships, other people's money, taxes, legacies and mortgages. What this all boils down to is the serious side of life and anything that involves uniting with others in marriage or business and then dealing with the finances or other serious matters on behalf of both you and your partner. Alternatively, you may deal with these things on a professional basis for others. You may find work in banking, accountancy or as a lawyer, where you have to keep an eye on money, business, land or possessions that belong to others.

You need to learn to balance your own requirements with those of others and to deal fairly and honestly with others and to be honest and fair to yourself. The second house South node may make you somewhat selfish when it comes to these vital and important elements of life. For instance, in a divorce situation, you may be the type who fights for the house, money, ownership of the children and even things that you don't really need or want.

Greed, slippery or sloppy behaviour in these areas will rebound on you, while too much sacrifice for the sake of others is no good either. There will be times when you will need to give financial or emotional support to others, but at other times you will need to call upon this for yourself. Don't become a martyr or you will allow resentment to build up and to destroy your relationships. There will be times in your life when you have to deal with the big issues of birth, death etc. and some of these may be quite harrowing.

The North Node in the Ninth house

This is another difficult house for a beginner to get to grips with as it rules such things as the law, education, religion, philosophy, truth, travel, foreigners and foreign goods. The central idea is expansion and you might expand your mind through education and travel and by getting to know people or dealing with goods from other countries. You can expand your intellectual and spiritual horizons through philosophy and religion, and you can test the boundaries of what can or cannot be done by dealing with ideas related to justice and legal matters.

Another central idea to this house is that of belief, so if your north node is here, you will spend time - perhaps years - exploring the world of ideas in order to find something that you can really believe in. The south node in the third house shows that you will have been given the tools by which you can read, write and learn and you will have been taught the beliefs of those who brought you up or educated you. The north node in the ninth house, forces you to explore your own ideas and to discover your own beliefs and to make clear judgements based on the facts.

The only real drawback here is that your striving for truth and honesty can make you a little too quick to offer your opinions, and you may upset others by expressing an honest opinion when a white lie would be more appropriate and less hurtful.

North Node in the Tenth house

The North node in the tenth house endows you with independence, strength of purpose and tremendous ambition. It may also give you delusions of grandeur! You won't want others to run your life but you may wish to run their lives, so the message here is to learn co-operation. Oddly enough the south node being in the fourth house should be a help because it indicates that you have brought the experience of living in a family with you from a previous incarnation.

You will definitely reach for a position of power and authority, and success may follow a period of struggle and hard work. However, when you finally become the political figure or the wealthy tycoon you would like to be, try to remember all those lesser mortals who need to rub along with you. The truth of this north node position is that you will probably have quite a hard furrow to plough, because you will not be born with a silver spoon in your mouth. You may fear returning to poverty and deprivation and you will do all you can to ensure future security for yourself and your family. The chances are that your family will always mean a lot to you and that all generations of them will be able to turn to you for help when they need it. Women with this north node will not be satisfied with a life of housework or playing second fiddle to a successful man, you will always need to do your own thing and to think your own thoughts.

North Node in the Eleventh house

The eleventh house is the place where detached relationships rule. This means friends, acquaintances and groups, committees, clubs and societies. If you have your north node here, you may become interested in causes or in doing things that make the world a better place for everybody, not just for yourself or your family. You will seek out those whose ideas run along the same lines as yourself or who share your passions and interests. There are two possible drawbacks to this house position. The first is that you may be too busy saving the whale to pay enough attention to those who love you and this may damage your personal relationships. The second problem is that you are so democratically minded that you may be satisfied with a low standard of achievement or personal fulfilment. The south node in the fifth house will remind you that you are a person in your own right and that you deserve respect and at least the same rights and benefits as the next person.

Friendship is a bit of a moveable feast with you, as you can become firm friends with one person or group of people and then move on easily to the next. This leaves others wondering whether you really liked them or approved of them in the first place. You may frustrate family members by almost forgetting that they exist while you are chasing your dreams.

This is also the house of eccentricity, so you may develop some really strange ideas. Try to keep a grip on reality and understand the necessity of living a normal emotional life and of giving real love to those who need it.

North Node in the Twelfth house

If you have the north node in this house you will have one foot in the next world. This means that you may be a natural psychic or medium and you will definitely be interested in everything related to these subjects.

There is a strong possibility that you will spend part of your life in some form of isolation. This may mean a lonely childhood, time spent being sick and unable to work, or in some other way detached from the hustle and bustle of daily life. If this does not occur, you may feel lonely within a family or within a marriage. Whatever the cause, the effect is that you will go on an inward journey to examine your beliefs and philosophy of life. Once you have come to terms with this, you may wish to work as a healer, teacher, medium or in some other way for the benefit of mankind. You may devote your life to those who are at the bottom of society's ladder of success.

Your challenge is not to sacrifice so much of yourself that you end up being drained. Even if you don't go out in to the wider world in order to help suffering humanity, you may become a doormat for a partner or for your family. You must learn to stand back from the urge to help just a little and fulfil your own needs first. You must rest, take time out and learn to say no from time to time or there will be nothing left.

You may become interested in the field of health, medicine or alternative therapies. Alternatively, you could become a hypochondriac or a person who moans and complains about their ailments. This will make you a thorough-going bore and it will damage or destroy your chances of making good relationships with others. In short, you need to learn to balance your self-involvement or your need to sacrifice all for the sake of others.

Chapter Ten:
Conjunctions to the Nodes

This chapter will make immediate sense to those of you who understand a little astrology, but it's perfectly easy for a beginner to follow.

When a planet or any other feature is close to another, this is called a conjunction. In the case of the nodes, the planet or feature in question must either be exact or at most a degree or two away from the node to be effective. Astrological theory suggests that it's best for something to contact the north node, but the fact is that a conjunction to either node works in the same way. When it comes to transiting or progressed planets triggering off events the situation is different, because transits to the north node are better than those to the south node. Having said this, I've discovered that some Oriental systems believe the exact opposite - so, check this out for yourself.

Conjunctions to the Nodes in a Birthchart

Sun, Moon, Ascendant or Midheaven
Success, fame, achievement and possibly wealth. Check out the sign and house to see how this manifests itself.

Mercury
The person will be a good speaker, writer or communicator and he may make a success of working in these fields. He may also succeed as a taxi-driver, mailman or bus driver who travels around his vicinity. Brothers, sisters, other relatives of around the same age and even neighbours will bring luck and benefits.

Venus
This is a good indicator for wealth. It also denotes an eye for beauty and symmetry that can lead to a career in any field that makes things look good. Music, the arts, fashion and cosmetics might appeal.

Jupiter

This conjunction brings luck in gambling and speculation, but it also has a lot to do with beliefs, so this person will search for and find a personal philosophy that makes his life interesting and this will also be of benefit to others. This person may make his fortune by travelling or by trading with people from other countries.

Saturn

Saturn is the planet of ambition, hard work and structure, so while this person's life may never be easy, he will make an effort at whatever he does and ultimately be very successful. The father, father figures or people in positions of authority will be helpful.

Uranus

Uranus is the planet of originality, so this person can make a success as an inventor or innovator and he can find unusual ways of doing necessary tasks. He may choose to work in electronics and other new technology in an innovative manner. Teaching will also appeal.

Neptune

Neptune is the planet of dreams, escapism and the imagination, so this person can turn his dreams, visions and ideas into reality. This is the sign of the artist and musician and also of one who explores psychic or mystical matters.

Pluto

This planet is often associated with medicine and psychology, so this person is bound to investigate these things. This powerful planet enables the person to transform his own personal life and perhaps to transform the lives of others.

Chiron

Chiron is associated with pain and healing, so this person can turn painful experiences to good account by understanding the pain of others and finding ways of healing them.

Other Planetary Aspects

This is not easy for a beginner to grasp, but it will make immediate sense to anybody with even a little knowledge of astrology. There are many other aspects, some of which are easy to live with and others which are challenging. If there is an aspect to one node, there will also be an aspect to the other.

- A sextile aspect, which is 60 degrees away from one node, will make a trine aspect, which is 120 degrees away from the other node. These are beneficial aspects.
- A square aspect will be 90 degrees away from both nodes, and this is an extremely challenging aspect.
- If a planet is 30 degrees away from one node, it will make a semi-sextile aspect and it will also be 150 degrees away from the other node, thus making an inconjunct. This is an irritating situation where luck can fluctuate between being mildly good in one area of life while causing difficulties in another.

Obviously, if one node is on the ascendant, the other will be on the descendant, which means that the person will be torn between independence and interaction with others. In this case, partnerships and relationships will bring both joy and tears.

When one node is on the midheaven and the other is on the nadir, this suggests that a person's career aims and ambitions may be at odds with his domestic and family life. In this case, the parents will be a strong influence for either good or ill.

Progressions and Transits

This is definitely beyond the scope of a non-astrologer, but those who know something about the subject will have no trouble understanding the following.

The planets will make aspects to the nodes by progression or transit at various times of a person's life. Some planets will make several aspects during the course of a year - and these aspects will coincide with or set off certain trends and events. This is taking us beyond the scope of this book, but a good astrologer will soon show you how these methods work and he will be able to explain the events that they predict.

Tradition states that a node cannot set off an event when it transits another planet because, "A node sheds no rays". The rays in question are not rays of light, but rays associated with the Qabalah and other spiritual traditions. However, tradition seems to forget that the ascendant, midheaven and so forth also shed no rays and they are certainly effective. Transiting nodes do create effects relating to all those lunar matters such as mother issues, household and domestic issues, the past, karma and so forth.

Once again, check these progressions and transits against your own chart and those of others, to see which node brings the most benefit, or the most trouble.

Chapter Eleven:
Eclipses

An eclipse occurs when the sun, moon and earth are in a line so that either the moon blocks the light of the sun (a solar eclipse) or when the earth prevents the sun's light from falling onto the moon (lunar eclipse). A solar eclipse occurs at the time of a new moon and a lunar eclipse occurs at the time of a full moon. The ecliptic is the path of the sun, so the moon must be right on that path for an eclipse to occur. We know that the nodes are the points where the moon crosses the ecliptic, so when an eclipse occurs, the nodes are always involved. A solar eclipse means that the sun, moon and one of the nodes are in the same place, while a lunar eclipse means that the sun is conjunct one node and the moon is conjunct the other. One type of eclipse is often (but not always) followed by a second two weeks later.

Harbingers of Doom

Ancient astrologers feared eclipses, and tradition has it that oracles foretold doom-laden events when an especially important eclipse was due. Like modern astrologers, these ancients knew that there are several partial eclipses every year and full solar or lunar eclipses every two or three years, so an eclipse had to fall on a sensitive area of a city or a nation's birthchart to have an effect. Although any planet or angle can be upset by an eclipse, the most sensitive points as far as eclipses are concerned are the sun and moon. These are the planets that are involved in the eclipse itself, so an eclipse will have a greater impact when it occurs on these "lights" than if they occur on an ascendant, midheaven or a planet, although they will be felt on any such spot. Modern astrologers are also aware of the eclipse effect, and we know that these planetary events can bring underlying discontent to the surface or enemies out into the open.

One famous eclipse prediction from ancient times was the destruction of Pompeii by the eruption of Vesuvius. Another from more recent times was the danger to America resulting from the eclipse in July 2001 which fell on the sun in the USA's chart.

When an eclipse falls on a personal chart, it will obviously have more impact if it falls on the sun, moon, a planet, the ascendant, midheaven, descendant, the nadir. However, it only needs to be at an angle to the sun or moon for it to have an impact. An eclipse can cause a devastating effect, which may even be life threatening in some circumstances. It brings problems to the surface in no uncertain terms.

One point to watch for is the period between two eclipses. For example, a friend of mine started what turned out to be a long and difficult relationship exactly half way between a solar eclipse and the lunar eclipse that followed a fortnight later, and another had a major operation exactly a week after a lunar eclipse and a week before the following solar eclipse.

Pre-natal Eclipses
The eclipse that took place just prior to your birth is called the pre-natal eclipse, and some astrologers research and use these. If you want this facility, you will need professional quality software. This will enable you to examine charts for people and places to see where these might have occurred both before the person or place came into being or just prior to a major event. A mapping chart will even be able to show you the part of the world that was most effected by an eclipse. Some astrologers see fame, fortune or extraordinary lives among those born shortly after an eclipse, especially if it's a lunar eclipse.

A Fresh Start
Despite the fact that eclipses on sensitive points in a chart are never easy, they do clear the way for a fresh start, rather like a boil that comes to a head and which can then be lanced. An eclipse rarely coincides with a new problem as it's usually a pre-existing one that comes into the open. Sometimes a situation that has been going on behind one's back comes out into the light. For those of

you who read the Tarot, an eclipse is rather like seeing the Tower in a Tarot reading.

Whether the effects of the eclipse are devastating or just mildly upsetting, they usually offer opportunity to clear the decks and make a fresh start. This may signal a time to call in a plumber to sort out an on-going practical problem or it may signify the need for a change of attitude. Whatever the scenario, an eclipse brings good as well as bad in its wake, because once the problem becomes obvious, it can be cleared away.

Ptolemy's Theorem
The ancient Egyptian astrologer Ptolemy had a theory about the timing of eclipses. This is how it works:

An eclipse on the ascendant will be felt
within three months of the eclipse.
On the nadir, three to six months later.
On the descendant, six to nine months later.
On the midheaven, nine months to a year later.

The signs of Cancer and Leo are particularly susceptible to being upset by eclipses, because Cancer is ruled by the moon, and Leo is ruled by the sun.

Table of Eclipses from 2000 to 2030
This table will enable you to check out the eclipses for some years past and into the future. It doesn't show whether the eclipses are total or partial, as this doesn't matter from the point of view of astrology. The first column tells you whether the eclipse is a solar or lunar one, the second shows gives the date, the third shows the time when the eclipse was exact, and the last column shows the degrees and minutes of the sign in which the eclipse occurs.

ABBREVIATIONS	
Ar:	Aries
Ta:	Taurus
Ge:	Gemini
Cn:	Cancer
Le:	Leo
Vi:	Virgo
Li:	Libra
Sc:	Scorpio
Sg:	Sagittarius
Cp:	Capricorn
Aq:	Aquarius
Pi:	Pisces.

TABLE OF ECLIPSES: 2000 - 2030

Type	Date	Time	Zone	Sign
Lunar	21 Jan 2000	04:40:26	UT	00°Le26'
Solar	05 Feb 2000	13:03:16	UT	16°Aq01'
Solar	01 Jul 2000	20:19:55	BST	10°Cn14'
Lunar	16 Jul 2000	14:55:13	BST	24°Cp19'
Solar	31 Jul 2000	03:25:09	BST	08°Le11'
Solar	25 Dec 2000	17:21:37	UT	04°Cp14'
Lunar	09 Jan 2001	20:24:24	UT	19°Cn39'
Solar	21 Jun 2001	12:57:45	BST	00°Cn10'
Lunar	05 Jul 2001	16:03:46	BST	13°Cp38'
Solar	14 Dec 2001	20:47:23	UT	22°Sg56'
Lunar	30 Dec 2001	10:40:32	UT	08°Cn47'
Lunar	26 May 2002	12:51:14	BST	05°Sg03'
Solar	11 Jun 2002	00:46:31	BST	19°Ge54'
Lunar	24 Jun 2002	22:42:21	BST	03°Cp11'
Lunar	20 Nov 2002	01:33:39	UT	27°Ta32'
Solar	04 Dec 2002	07:34:21	UT	11°Sg58'
Lunar	16 May 2003	04:35:57	BST	24°Sc52'
Solar	31 May 2003	05:19:52	BST	09°Ge19'
Lunar	09 Nov 2003	01:13:21	UT	16°Ta12'
Solar	23 Nov 2003	22:58:56	UT	01°Sg14'
Solar	19 Apr 2004	14:21:11	BST	29°Ar49'
Lunar	04 May 2004	21:33:26	BST	14°Sc41'
Solar	14 Oct 2004	03:48:15	BST	21°Li06'
Lunar	28 Oct 2004	04:07:22	BST	05°Ta02'
Solar	08 Apr 2005	21:32:00	BST	19°Ar05'
Lunar	24 Apr 2005	11:06:28	BST	04°Sc19'
Solar	03 Oct 2005	11:27:52	BST	10°Li18'
Lunar	17 Oct 2005	13:13:38	BST	24°Ar13'
Lunar	14 Mar 2006	23:35:26	UT	24°Vi14'
Solar	29 Mar 2006	11:15:14	BST	08°Ar34'
Lunar	07 Sep 2006	19:42:03	BST	15°Pi00'
Solar	22 Sep 2006	12:45:03	BST	29°Vi20'
Lunar	03 Mar 2007	23:17:06	UT	12°Vi59'

Solar	19 Mar 2007	02:42:33	UT	28°Pi07'
Lunar	28 Aug 2007	11:35:05	BST	04°Pi45'
Solar	11 Sep 2007	13:44:13	BST	18°Vi24'
Solar	07 Feb 2008	03:44:30	UT	17°Aq44'
Lunar	21 Feb 2008	03:30:30	UT	01°Vi52'
Solar	01 Aug 2008	11:12:33	BST	09°Le31'
Lunar	16 Aug 2008	22:16:27	BST	24°Aq21'
Solar	26 Jan 2009	07:55:16	UT	06°Aq29'
Lunar	09 Feb 2009	14:49:09	UT	20°Le59'
Lunar	07 Jul 2009	10:21:25	BST	15°Cp24'
Solar	22 Jul 2009	03:34:36	BST	29°Cn26'
Lunar	06 Aug 2009	01:54:52	BST	13°Aq43'
Lunar	31 Dec 2009	19:12:45	UT	10°Cn14'
Solar	15 Jan 2010	07:11:22	UT	25°Cp01'
Lunar	26 Jun 2010	12:30:21	BST	04°Cp46'
Solar	11 Jul 2010	20:40:27	BST	19°Cn23'
Lunar	21 Dec 2010	08:13:27	UT	29°Ge20'
Solar	04 Jan 2011	09:02:36	UT	13°Cp38'
Solar	01 Jun 2011	22:02:36	BST	11°Ge01'
Lunar	15 Jun 2011	21:13:34	BST	24°Sg23'
Solar	01 Jul 2011	09:53:54	BST	09°Cn12'
Solar	25 Nov 2011	06:09:40	UT	02°Sg36'
Lunar	10 Dec 2011	14:36:22	UT	18°Ge10'
Solar	21 May 2012	00:47:00	BST	00°Ge20'
Lunar	04 Jun 2012	12:11:33	BST	14°Sg13'
Solar	13 Nov 2012	22:07:59	UT	21°Sc56'
Lunar	28 Nov 2012	14:45:55	UT	06°Ge46'
Lunar	25 Apr 2013	20:57:06	BST	05°Sc45'
Solar	10 May 2013	01:28:22	BST	19°Ta31'
Lunar	25 May 2013	05:24:55	BST	04°Sg08'
Lunar	19 Oct 2013	00:37:39	BST	25°Ar45'
Solar	03 Nov 2013	12:49:56	UT	11°Sc15'
Lunar	15 Apr 2014	08:42:16	BST	25°Li15'
Solar	29 Apr 2014	07:14:19	BST	08°Ta51'

Lunar	08 Oct 2014	11:50:34	BST	15°Ar05'
Solar	23 Oct 2014	22:56:38	BST	00°Sc24'
Solar	20 Mar 2015	09:36:08	UT	29°Pi27'
Lunar	04 Apr 2015	13:05:31	BST	14°Li24'
Solar	13 Sep 2015	07:41:13	BST	20°Vi10'
Lunar	28 Sep 2015	03:50:28	BST	04°Ar40'
Solar	09 Mar 2016	01:54:25	UT	18°Pi55'
Lunar	23 Mar 2016	12:00:46	UT	03°Li17'
Lunar	18 Aug 2016	10:26:30	BST	25°Aq51'
Solar	01 Sep 2016	10:03:02	BST	09°Vi21'
Lunar	16 Sep 2016	20:05:02	BST	24°Pi20'
Lunar	11 Feb 2017	00:32:48	UT	22°Le28'
Solar	26 Feb 2017	14:58:18	UT	08°Pi12'
Lunar	07 Aug 2017	19:10:32	BST	15°Aq25'
Solar	21 Aug 2017	19:30:06	BST	28°Le53'
Lunar	31 Jan 2018	13:26:38	UT	11°Le37'
Solar	15 Feb 2018	21:05:06	UT	27°Aq07'
Solar	13 Jul 2018	03:47:46	BST	20°Cn41'
Lunar	27 Jul 2018	21:20:15	BST	04°Aq44'
Solar	11 Aug 2018	10:57:38	BST	18°Le41'
Solar	06 Jan 2019	01:28:04	UT	15°Cp25'
Lunar	21 Jan 2019	05:15:57	UT	00°Le51'
Solar	02 Jul 2019	20:16:05	BST	10°Cn37'
Lunar	16 Jul 2019	22:38:05	BST	24°Cp04'
Solar	26 Dec 2019	05:13:00	UT	04°Cp06'
Lunar	10 Jan 2020	19:21:09	UT	20°Cn00'
Lunar	05 Jun 2020	20:12:14	BST	15°Sg34'
Solar	21 Jun 2020	07:41:18	BST	00°Cn21'
Lunar	05 Jul 2020	05:44:16	BST	13°Cp37'
Lunar	30 Nov 2020	09:29:32	UT	08°Ge38'
Solar	14 Dec 2020	16:16:26	UT	23°Sg08'
Lunar	26 May 2021	11:13:42	UT	05°Sg25'
Solar	10 Jun 2021	10:52:28	UT	19°Ge47'

Lunar	19 Nov 2021	08:57:17	UT	27°Ta14'
Solar	04 Dec 2021	07:42:51	UT	12°Sg22'
Solar	30 Apr 2022	20:27:54	UT	10°Ta28'
Lunar	16 May 2022	04:13:57	UT	25°Sc17'
Solar	25 Oct 2022	10:48:30	UT	02°Sc00'
Lunar	08 Nov 2022	11:01:57	UT	16°Ta00'
Solar	20 Apr 2023	04:12:18	UT	29°Ar50'
Lunar	05 May 2023	17:33:50	UT	14°Sc58'
Solar	14 Oct 2023	17:54:55	UT	21°Li07'
Lunar	28 Oct 2023	20:23:49	UT	05°Ta09'
Lunar	25 Mar 2024	07:00:04	UT	05°Li07'
Solar	08 Apr 2024	18:20:36	UT	19°Ar24'
Lunar	18 Sep 2024	02:34:13	UT	25°Pi40'
Solar	02 Oct 2024	18:49:02	UT	10°Li03'
Lunar	14 Mar 2025	06:54:22	UT	23°Vi56'
Solar	29 Mar 2025	10:57:33	UT	09°Ar00'
Lunar	07 Sep 2025	18:08:37	UT	15°Pi22'
Solar	21 Sep 2025	19:53:51	UT	29°Vi05'
Solar	17 Feb 2026	12:00:51	UT	28°Aq49'
Lunar	03 Mar 2026	11:37:35	UT	12°Vi53'
Solar	12 Aug 2026	17:36:27	UT	20°Le02'
Lunar	28 Aug 2026	04:18:14	UT	04°Pi54'
Solar	06 Feb 2027	15:55:47	UT	17°Aq37'
Lunar	20 Feb 2027	23:23:19	UT	02°Vi05'
Lunar	18 Jul 2027	15:44:36	UT	25°Cp48'
Solar	02 Aug 2027	10:04:54	UT	09°Le55'
Lunar	17 Aug 2027	07:28:22	UT	24°Aq11'
Lunar	12 Jan 2028	04:02:44	UT	21°Cn27'
Solar	26 Jan 2028	15:12:10	UT	06°Aq10'
Lunar	06 Jul 2028	18:10:28	UT	15°Cp11'
Solar	22 Jul 2028	03:01:22	UT	29°Cn50'
Lunar	31 Dec 2028	16:48:10	UT	10°Cn32'
Solar	14 Jan 2029	17:24:08	UT	24°Cp50'
Solar	12 Jun 2029	03:50:10	UT	21°Ge29'
Lunar	26 Jun 2029	03:21:58	UT	04°Cp49'
Solar	11 Jul 2029	15:50:41	UT	19°Cn37'
Solar	05 Dec 2029	14:51:45	UT	13°Sg45'
Lunar	20 Dec 2029	22:46:08	UT	29°Ge20'

Chapter Twelve:
Solar Eclipses

A solar eclipse occurs when the sun and moon are in the same sign and when the moon blocks the light of the sun from the earth.

An Eclipse in the Element of Fire
Masculine signs
Aries, Leo, Sagittarius

If an eclipse falls on a sensitive point in a fire sign, the energies that are released are impulsive, enthusiastic and sometimes aggressive, and there will be feelings of passion, ardour and impatience, which can be useful. This is the time to bring your feelings out into the open rather than repressing them. If anger and resentment are continually stuffed down and never expressed, they will eventually erupt in sickness, an overpowering desire for sleep or a level of emotional pain that barely permits you to function normally. It's important at such a time to force your feelings into the open through talk, action and perhaps through strenuous physical activity. If you feel really blocked or frustrated, try throwing a tantrum in a locked room with loud music blaring!

Under the influence of fire, your instinctive responses are spontaneous and you will find the vitality needed to get new issues off the ground. You may actually discover extra joy in the little things in life, although this period of your life is more likely to symbolize the big things, the grand gestures and the great events. The effect of a fiery eclipse may sweep you away in the drama of an impulsive affair that reek of excitement and instability. Your feelings may flood out in a rush of emotion that leaves you no time to think and you may love more passionately and intensely, but the chances are that an affair started at such a time will not last.

If you are in the habit of restraining your feelings, your behaviour will come as a shock to those who are around you because suddenly now your temper will feel like a blazing torch on a dark night. Although minutes after the initial blaze you will forget your anger or turn it elsewhere. This is exactly the time when you may feel an urge for excitement and an impulse to court danger. You will be challenged by your experiences and find that you are more interested in the pursuit than the ultimate pleasure. If you can walk away from a tempting situation you may save yourself from pain, but you may also miss out on an exciting and memorable experience.

An Eclipse in the Element of Earth
Feminine signs
Taurus, Virgo, Capricorn

An eclipse in earth signs brings pragmatic, materialistic productive and logical energies to the fore. This will enhance your ability to persevere with problems, discipline your mind and adopt a patient attitude towards life in general. An earthy eclipse will energize you to take life one step at a time, to avoid acting on impulse and wait patiently for larger rewards. You may be forced to take on a tough task or one that you know will take time to complete. Earthy eclipses emphasize hard work, paying attention to tedious details and organising your life and work to operate with maximum efficiency.

Material possessions, negotiating money matters and an instinct for making money grow can come under this influence. Here is where creature comforts over-ride principles – the thought of starving or living in poverty is not an option when the element of earth is involved. If the eclipse has an upsetting effect, it will be your finances, possessions and perhaps even your land and property that will suffer at this time. An upsetting event may force you to focus strongly on your goals and you will attack everything in your life with persistence and energy.

The earth element emphasizes the luxuries of life – not in a frivolous way, but as in investment in the kind of art or jewellery that gives pleasure for years. The downside is that you may lose something of this nature, but you may be able to make use of the eclipse influence to spot the things that will make a profit, and you may capitalize on opportunities. Either way, if you know that such an eclipse is coming up, it would be worth taking out extra insurance cover!

In emotional matters, what you now seek is an opportunity to demonstrate loyalty responsibility and steadfastness whilst you work towards emotional and material security. The chances are that any relationship that you enter into at this time may promise these things but not actually deliver them. Alternatively, you may realize that a current relationship is not going in this direction. As always, the eclipse brings issues to a head and gives you an opportunity to review a situation. Game playing, emotional immaturity and ambivalence will irritate you more than usual. Use the influence of the earth to overcome any tendency you might have to get into or remain in a relationship for all the wrong reasons.

An Eclipse in the Element of Air
Masculine signs
Gemini, Libra, Aquarius

An eclipse on a sensitive point in an air sign will focus on mental and intellectual rather than emotional matters. Experiences are analysed rather than felt by the heart and you will rely more heavily on intellect than intuition or emotions. The drawback is that you may realize that something you are studying or working on is going nowhere, and that you need to scrap your original plan and think up a new one. Another downside is that a friend may let you down. Friendships may drift away or even come to a rather nasty end at this time. New friends will soon come into view now and there will eventually be fresh in social events to enjoy.

On one hand, this may be a time of arguments and even of verbal abuse. If you feel like abusing someone else, try to hold your tongue, because once the eclipse is over whatever set you off may vanish into the mists, leaving your friend or loved one with painful memories. On the other hand, a measure of intelligence and emotional objectivity will come under the spotlight of an airy eclipse, and you can tap into this to analyse a situation and see whether it's still viable or not. Airy influences can bring a desire for endless discussion about feelings, verbal exploration into the meanings of situations and analysis of those who are closely involved in relationships with you. You will make a strong effort to understand others and this may encourage tolerance of people and situations.

Eclipses and influences in air signs enable you to be more adaptable, versatile and perhaps more than usually restless. This situation may be forced upon you by dint of circumstances, or you may just be badly in need of a change of view. You may wish to detach from those around you and lose yourself in reading and writing. Alternatively, you may find it easier to talk with outsiders than your loved ones. You will be happy to live on the surface of life for a while at this time. This is fine for a while (and it may be needed), but it shouldn't be allowed to go on for too long. Your feelings will be ambivalent and you may not be sure what you want from a relationship or from your career. You may feel an urge to travel, move house, buy a vehicle or gain intellectual or mental freedom.

The influence of an air sign eclipse may encourage you to waste precious time in pipe dreams that have no hope of materializing into anything concrete. You may find that your plans never get past the thinking stage or that you change your mind in midstream. This eclipse offers an opportunity for you to positively nurture a greater sense of realism along with some firm discipline. If you use it to learn to look at both sides of a situation before you act, you will take advantage of its beneficial effects and then give life to original ideas.

An Eclipse in the Element of Water

Feminine signs

Cancer, Scorpio, Pisces

This is where emotions, intuition, sensitivity and vulnerability are highlighted. You may have spent years hiding your feelings and you may have built up an array of defences in order to minimize hurt. The problem is that these defences will now be blown away, leaving you to face reality. Deep feelings that are related to past emotional experiences will emerge and you may discover a tendency to bear grudges for offences committed against you. Even if you think you have evolved beyond this, you won't find it easy to forgive and forget when a watery eclipse occurs.

You will find it easy to use persuasion, intelligence and a razor sharp memory at this time, and you will be determined to understand what is going on around you. This may relate to matters that affect you on several levels – from the people in your life to your philosophical outlook. The universe may suddenly appear to be a fascinating mystery which titillates you and which enables you to discover meanings you had never considered before. A water sign eclipse may enhance an immediate natural understanding of other people's motives and an ability to see things from another's point of view.

Your imagination will take off and creative projects that require imagination and will occupy your mind. Eclipses being what they are, you may write a novel, only to have it turned down by a publisher. The answer here is to look at your work, rewrite it if necessary and find another publisher. If you can put your mind to new projects and find the necessary self-discipline to continue once your passion and enthusiasm has passed the planning stage, you could go on to make a real success.

The influence of water brings to the fore a quest for emotional security through an intense loving relationship that will sustain the initial passion. Indeed, a sudden loss or reversal in a relationship may force you to recognize this need as crucial to your life. As with all eclipses, there is a danger of being tempted by the thrill of superficial excitement that appeals to the lower levels of sensuality, such as drug or alcohol abuse. The highest level of water energy allows you to seek profound and enduring spiritual satisfaction.

Chapter Thirteen:
Lunar Eclipses

A lunar eclipse occurs when the moon is full, so the earth shades it from the light of the sun. When two planets are on either side of the earth on a horoscope they are in *opposition* to each other.

The sun rules the personality and it represents those things that you initiate and the way that you operate. The moon rules the hidden, vulnerable, romantic and emotional side of your nature, and it may be a far better gauge of what you are groping for on an instinctive level. When a lunar eclipse occurs, the sun is in one sign and the moon in an altogether different one, revealing the inner conflict between your conscious and unconscious self that may come to the surface at the this time.

The signs of the zodiac are linked so that the *fire* signs always oppose *air* signs and vice versa, while *earth* oppose *water* signs and vice versa. The table and illustration below will show the pattern:

SIGN & ELEMENT	OPPOSITE SIGN & ELEMENT
Aries / fire	Libra / air
Taurus / earth	Scorpio / water
Gemini / air	Sagittarius / fire
Cancer / water	Capricorn / earth
Leo / fire	Aquarius / air
Virgo / earth	Pisces / water
Libra / air	Aries / fire
Scorpio / water	Taurus / earth
Sagittarius / fire	Gemini / air
Capricorn / earth	Cancer / water
Aquarius / air	Leo / fire
Pisces / water	Virgo / earth

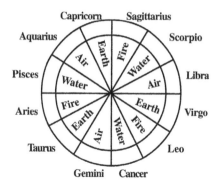

The Elements of the signs

Fire and Air Combinations

Aries / Libra
Leo / Aquarius
Sagittarius / Gemini
Gemini / Sagittarius
Libra / Aries
Aquarius / Leo

You will want to shine, to take charge, to take precedence and to be the centre of attention, but you will also want to stand back, analyse the situation and ask yourself whether you really want this after all. The effects of this eclipse may make you rush forward at one moment, then hold back and ask yourself, your partner or your friends whether you are doing the right thing. You will have great ideas, but you may spoil things by displaying too much ego, pride, temper or selfishness. You need to discuss your feelings and future expectations openly. As long as your ego does not feel compromised or threatened, such discussion will be a successful way of communicating. Beware of over analysing your motivation and remember that you have a need to maintain personal freedom in order to remain in control. Conflicts and clashes can only be resolved by emotional maturity.

Nothing happens slowly when fire or air are in effect, so you can expect domestic, business or relationship problems to come to the surface in a very immediate and obvious way. This may be the time when you realize that you simply cannot put up with a bullying mother-in-law for a moment longer or that you simply have to move house. It will take several months, perhaps even a year or two before you can put the change into action, but the eclipse effect will bring a sudden and very real moment of revelation. A friend of mine told me that the reality of her abusive marriage came home to her on the day when her husband punched her in the face and threw her down the stairs. It took a few more years before she was finally able to leave this man, but it was a lunar eclipse that fell exactly on her sun in a fire sign that brought this realization into the light.

Earth and Water Combinations
Taurus / Scorpio
Virgo / Pisces
Capricorn / Cancer
Cancer / Capricorn
Scorpio / Taurus
Pisces / Virgo

The effect here is to bring confusion to practical matters or alternatively to damp down feelings, intuition and emotion to the point where you are not sure if what you are feeling is right. Someone around you may have needs that you simply can't tap into at this time, or they may be elated or unhappy for reasons that you just can't pick up on. Alternatively, you may well know exactly what is going on but be too tired to cope with the demands and problems of others while being unable to avoid getting dragged into them. You may have no time for different points of view and you may not agree with those who seek to give you their opinions, although if you set aside your usual opinions you could find yourself enjoying an exchange of ideas and stories. You may have to make an effort to give someone else personal space, and by doing this you could lay the foundation for years of worthwhile friendship or a close relationship that endures.

You may seek emotional security but be temporarily unable to find it. During this eclipse period you will need to show patience and understanding towards yourself as well as towards others. Your moods may fluctuate wildly for a while. You will have a lot to offer at this time, as your mood is loving, sensitive, giving and profound. The energies of this eclipse may enable you to off-load a relationship that's doing you harm and ultimately to find one that gives you the love, trust and security that you long for. Alternatively, you may have to re-examine your own behaviour and perhaps to rise above your own needs in order to achieve a closer emotional relationship with the person you already have.

An unexpected domestic disaster of a practical nature may take up your time, so that you spend hours trying to get something fixed in the home, on your vehicle or at work. You may be so angry, upset or even frightened that you can't see how to get the problem fixed or sorted out. The only thing to do is to sit down, make a cup of your favourite brew and take a few deep breaths. Better still, if the problem can be left for a day or two until the eclipse has passed, you will probably find the solution. A serious problem at work may make you re-think your work methods in order to become more efficient or productive. You may even have to consider your true aims and ambitions and perhaps look for a new career or even to take some time out and go back to school in order to gain a specific qualification.

Chapter Fourteen:
Eclipse Aspects

f you know the exact degree of your planets, you can check to see whether an eclipse makes an aspect to any of them.

The Aspects

Zero degrees - the conjunction
Here, the eclipse is on a planet. We have already looked at this in the section on solar eclipses in this chapter.

180 degrees - the opposition
This is the case when a solar eclipse opposes a planet. The effect of this is similar to the conjunction, but it frequently brings relationships with others into prominence.

Zero degrees and 180 degrees - the conjunction and opposition
This occurs when a lunar eclipse puts either the sun or your planet and the moon opposite it or vice versa. We have talked about lunar eclipses already in this chapter.

30 degrees - the semi-sextile
When a solar eclipse happens in the sign adjacent to one of your planets. The effect is so mild as to be almost unnoticeable.

150 degrees - the inconjunct
This is when a solar eclipse occurs five signs away from your planet. This is an irritating aspect which may hit your health, your job, your relationships or your finances and joint emotional or financial matters. There may be something wrong with your sex life or you may be unhappy about a matter related to fertility and conception.

30 degrees and 150 degrees - the semi-sextile and the inconjunct
A double-whammy, when a lunar eclipse hits both these points at once. The effect is like the inconjunct but more irritating, as

it seems to affect health, work, finances, possessions, your self-image and self-confidence.

60 degrees - the sextile

This is where a solar eclipse makes an aspect that is two signs away from the planet. It's interesting, enlightening, mildly irritating or somewhat beneficial.

120 degrees - the trine

Here we have a solar eclipse four signs away from your planet. This is irritating, but also enlightening and possibly useful if you are struggling to see what is wrong with a creative project, or if there is a problem related to younger members of your family. This can bring a short-lived love affair to an end.

60 degrees and 120 degrees - the sextile and trine

This is where a lunar eclipse hits these two points at once while making an aspect to your planet. This can be enlightening and slightly irritating. It can make you very busy, under pressure and a bit muddle-headed for a while. It can point out the areas of difficulty in a creative project or bring a short-term love affair to an end. This eclipse can even be mildly pleasant.

90 degrees - the square

This is where a solar eclipse is three signs away from your planet. This is pretty nasty and it may make life as difficult an eclipse in conjunction with a planet. You will definitely face challenges at this time and things that come to light now may not be pleasant. Your self-confidence could take a knock, business matters may be difficult, a child may play up or a creative venture could suddenly fail or need to be radically altered.

90 degrees each side - the "T" square

This is an aspect where a lunar eclipse forms a "T" square with your planet. It's likely to be extremely unpleasant to live through and it can be as bad as or worse than an eclipse that conjuncts or opposes a planet. This could affect family and domestic life, your career, your aspirations, issues relating to your parents or authority figures and older people in general.

Chapter Fifteen:
Occultations

The word *occultation* comes from the same root as occult, meaning hidden, secret or tucked away from sight. When something is occluded, it's hidden from view, while an occlusion is something that you can't see just by looking at it. Your dentist will know all about occlusions as these are holes that are tucked inside a tooth. Very few astrologers take occultations into account - for the simple and rather amazing reason that they haven't a clue what the symbols mean or how to interpret them, so you will soon be one jump ahead of the rest!

Planetary Occultations
There are many occasions when the moon passes in front of a planet in the solar system, and such occultations definitely have an effect on us. Astronomers and astrologers know exactly where the moon will be at any time, so these occultations are easily predicted.

Although what I am going to describe here is extremely easy to understand, it will initially make far more sense to those who already know something about astrology than it will to a complete beginner. You may wish to watch the sky to see an occultation in action, but the chances are that you will simply look them up in a set of tables that is called an astrological ephemeris. The particular ephemeris that you will need for this purpose is the annual "Raphael's Astronomical Ephemeris". Find Raphael's Ephemeris on Amazon.co.uk/com. I don't have a current price for it, but it's not expensive.

The following image is a typical page from Raphael's Ephemeris showing the planetary positions, and in this case, an occultation of Saturn by the moon on the 9th of December.

Lunar Aspects
Open the ephemeris at any page that shows the transits of the planets and look at the batch of columns on the far right hand side, which is marked Lunar Aspects. This column shows the aspects that the moon makes to each of the planets, starting with the sun on

EPHEMERIS				DECEMBER												25
D	☿	♀	♂	♃	♄	♅	♆	♇	Lunar Aspects							
M	Long.	Long.	Long.	Long.	Long.	Long.	Long.	Long.	☉	☿	♀	♂	♃	♄	♅	♆	♇
1	0♑35	23♐19	16♐58	16≈38	13♈44	5≈42	27♐57	5♐37		♂	⊼		∠		∠	⊼	
2	1 21	24 3	17 44	16 47	13R 43	5 44	27 59	5 40	⊼				□	⊼			⊼
3	2 1	24 45	18 31	16 57	13 41	5 46	28 0	5 42	∠		♂	♂			♂		∠
4	2 33	25 26	19 17	17 6	13 40	5 49	28 2	5 44	⚹⊼	⊼			⚹	♂			⚹
5	2 58	26 6	20 4	17 16	13 39	5 51	28 4	5 47	∠	⊼	⊼	♂					
6	3 15	26 45	20 50	17 26	13 37	5 53	28 6	5 49	⚹		∠		∠	⊼	⊼		□
7	3 22	27 22	21 37	17 36	13 36	5 56	28 7	5 51	□		∠	⚹	⊼	∠	∠		
8	3R 19	27 59	22 23	17 46	13 36	5 59	28 9	5 54		□	⚹		∠		⚹	⚹	△
9	3 5	28 34	23 10	17 57	13 35	6 1	28 11	5 56	△			□	⚹	•			△
10	2 40	29 7	23 56	18 7	13 34	6 4	28 13	5 58	⚼	△	□				□	□	
11	2 3	29♐39	24 43	18 17	13 33	6 6	28 15	6 1		⚼		□	⊼				
12	1 15	0≈ 9	25 30	18 28	13 33	6 9	28 17	6 3			△	△	∠	△	△		⚼
13	0♐16	0 38	26 16	18 39	13 33	6 12	28 18	6 5		⚼	⚼	△	⚹		⚼		
14	29♐ 9	1 5	27 3	18 50	13 32	6 14	28 20	6 8	⚼	⚼			⚼		⚼		
15	27 53	1 30	27 50	19 1	13 32	6 17	28 22	6 10						□			

the left and ending with Pluto on the far right. In this ephemeris, Pluto is still expressed with the old glyph, which is a combination of the letters P & L, rather than the more modern one that looks like the glyph for Mercury with a small circle on top.

When you look down the columns, you will see the aspects that the moon makes to the planets. You can check out what kind of aspects all these glyphs symbolize by referring to the key at the beginning of the ephemeris, but for now, we are only concerned with two of these symbols, namely the conjunction and the opposition.

The Conjunction   **The Opposition**

Once in a while, you will see the same symbols, but with the circles filled in completely rather than being empty. The filled-in conjunction symbol that you find in the column for lunar aspects to the sun indicates a solar eclipse, and a filled-in opposition symbol in the column for the sun indicates a lunar eclipse.

The only filled-in symbol that you will ever see in the columns for any of the planets other than the sun is the one for the conjunction. This indicates that the moon will pass in front of the planet in question at some point during the course of that day, and this will hide or occlude the planet - hence this is an occultation. The moon moves at a rate of half a degree per hour, so it can appear to snuff out a distant star within anything from a few minutes to an hour, but the planets in our solar system are not that far away so an occultation will take as much as a couple of hours to clear. It's the planetary occultations that we will now look at - because for one thing, the ephemeris tells us when they will occur.

Months go by with no occultations, and then they bunch up, often within days of one another, often selecting one or two slow-moving planets for special treatment. This is because the planets in question are lying in the right spot along the moon's course at that time.

Stellar Occultations

It's just possible that an astrologer who is interested in the distant "fixed" stars and who has the patience to discover when the moon will occlude one of them would be able to interpret these events astrologically. This is not impossible to do, because modern astrological software gives the longitude of the fixed stars that lie along the ecliptic, so as long as the moon is in the same position along this line and at the same declination (height), above or below the line as the star, this can be plotted.

Have you ever looked up in the sky on a dark night and watched the moon as it swallows a star? This is what is called an occultation and it occurs when the moon travels across the sky and passes in front of a star. For more than two centuries, observers - both amateur and professional - have recorded apparently impossible variations on a typical occultation. Instead of being completely obscured by the passing moon, occluded stars occasionally have been seen to remain fixed in the sky, and they can look as if they are hanging on the moon's limb for a second or so. On very rare occasions, a star has actually appeared to pass in front of the moon for a few seconds. This is obviously an optical illusion, as the moon is only quarter of a million miles away from us, while the closest star to us other than our own sun is Alpha Centauri Proxima, which is four and a half light years distant.

An occultation can last for as long as an hour or be over in a few minutes. The effect is immediate, so like a candle that's snuffed out, the star suddenly disappears and then just as suddenly re-emerges from the other side of the moon and blinks on again. But an occultation does not always block the light of a star completely. When the moon is not full, the starlight remains visible. In "grazing" occultations, the star seems to blink on and off as it crosses the edge of the moon as for a brief moment the star's light is blocked by the peaks on the moon's surface. Occultations demonstrate to us that the moon has no atmosphere. If there was a lunar atmosphere, stars would not seem to be suddenly snuffed out when they are occluded by the moon.

An observer recorded in 1783 that Sir William Herschel was conducting an observation when a companion placed herself at the telescope and watched a star disappear. Suddenly, she saw it reappear in front of the moon rather than behind it. Herschel stepped up to the telescope, saw a bright point on the dark disc of the moon and followed it carefully, watched it becoming fainter and fainter, until it finally vanished. Other observers over the years have reported similar phenomena. Various reports have ranged from seeing a star that emerged at the moon's dark limb and hung there for a few seconds, to one that was suspended for so long that the observers got tired of waiting.

In 1928, an observer in Scotland was so mystified by an occultation that he reported a sight he was convinced he would never forget. After the moon and star crossed paths, the light of the star did not just remain on the edge of the moon for a few seconds. It grazed about two degrees past the edge as if the star was between the moon and the earth, then it began to diminish until it finally vanished. The famous British astronomer, Patrick Moore, witnessed a strange occultation in 1972 together with a colleague who verified his report. The star he was watching faded out slowly instead of vanishing suddenly. Moore later explained that this could be explained by the fact that the star was a binary – a system made up of two stars which were so close together that they appeared as one. The "fading" was due to the fact that the two stars were hidden at fractionally different times. It's likely that any instances of such fading occultations are due to the presence of a binary system.

Chapter Sixteen: Interpreting the Occultations

An occultation of a planet takes an hour or two to pass, but its effects are often out of all proportion to the time that it takes. If a planet happens to lie on the path of the moon for a period of time, this will trigger an occultation every four weeks. If your life is in turmoil for a few months, it may be an occultation that's responsible for such a series of strikes at your nerves and emotions. As with an eclipse, an occultation won't necessarily bring a new situation into being, but it will make you aware of a current one and perhaps bring something out into the open. Most frequently, an occultation ends a situation, or it warns that one needs to end.

The moon refers to feelings and emotions, domestic or family circumstances, so the occultation may affect your home life, family life and your emotions. However, the planet that is occluded will have its own agenda and that's what the moon will block. The interpretations below will show you the kinds of events that might occur when a particular planet is occluded by the moon.

The Occultations of the Planets

The Moon Occluding Mercury
This can make you feel frustrated, annoyed, depressed, muddled or unsure of yourself. Your brain may not function with its usual speed and accuracy and you may feel quite stupid for a while. There may be a problem in the home or the family in connection with brothers, sisters, step-relatives or cousins of your own generation or with neighbours.

Your vehicle may let you down or a bus or train might be cancelled when you most need it. You may go shopping with a specific thing in mind and come home with everything but the one thing you went out for. Alternatively, you could run around doing errands but forget the most important of them. Even if your brain is working well enough for you to remember the errand, the place may be unexpectedly closed. If you rely upon telephones, mobile cell-phones or some other form of communications equipment, they may temporarily go on the blink or you could forget to make an important call. There may be a power cut, or worse still a power surge that affects your computer. Something related to your work could suddenly go wrong and any creative or craft hobby suddenly could lose its appeal.

You may open your mouth and find exactly the wrong thing coming out of it, or you may be unexpectedly lost for words. Alternatively, you may just be overtired and under the weather.

Venus

An occultation of Venus can have an amazingly powerful effect on relationship matters. It can signal the end of a relationship or it can coincide with a blazing row. Someone who has hitherto remained a secret enemy can suddenly come out into the open, or someone who has been secretly undermining you can suddenly be seen for the traitor they really are. Feelings of jealousy, rage, disappointment and loss can suddenly surface. One specific woman, or women in general, can upset you badly at this time.

I remember one acquaintance who had been seeing a man who had assured her that he had finished with his previous girlfriend, but this turned out not to be the case. My friend had to endure jealous and unpleasant phone calls from this woman until one day she woke up to the fact that the man was still seeing the previous girl and that he had lied to her all down the line. She rapidly brought the affair to an end and got both of them out of her life. The row that she had with the boyfriend occurred right on a Venus occultation, and true to form, the final confrontation took place in her home (remember: the moon rules the home).

Your self-esteem might take a knock and you may behave in a way that is beneath you. If you go out on a special date or for a special event, you may be unhappy with your hair or with the outfit you have chosen, or someone may upset you and spoil the mood of the occasion. You may feel fat, ugly, unfashionable or just fed up with yourself. In this case, the occultation is probably telling you to take yourself and your wardrobe in hand.

There may be some kind of financial loss or perhaps something in your home will suddenly go wrong, thereby causing expense and inconvenience. The usual culprits when the moon is involved are the fridge or cooker, but Venus rules mirrors, Venetian blinds and objects of beauty or value, so any of these can be damaged. It may even be worth checking your household insurance policy when you see such an occultation looming on the horizon.

Mars

This occultation will do nothing for your temper (unless you really need to blow off steam) because you will definitely feel like exploding. The lunar connection might point to your relationship with a mother or mother figure being the one to set you off on this tack, but it could also be a child and your own position as mother that's at the heart of this matter. The Mars connection could signify that a man lets you down or that one quite unexpectedly picks a fight with you. If you need to deal with professional men, workmen and so forth, they will be more trouble than they are worth at this time.

Guard against accidents when an occultation of Mars is in the offing, as you may cut or burn yourself, or you may break something that's useful or important to you. Take care when handling machinery and especially when driving or riding, but be especially careful in and around your home. Fire is a real risk at this time, so take care with anything that could cause this, and equip your home with smoke alarms, extinguishers and fire blankets just in case.

Jupiter

This could denote a need to change mind about something that you hitherto believed in. Something could occur that makes you realize that your thinking on some important factor was wrong, but it could equally signal the fact that you are right and that those who are around you are wrong.

There are a number of other weird and wonderful possibilities that could occur on this occultation. This is not a good time to take chances or to gamble or speculate on a business matter or any kind of chancy enterprise. Matters related to education or some other means of expanding your horizons could suddenly go wrong for a while. There may be a legal or official setback or travel plans may become fouled up. There may even be a problem related to any animals that you are responsible for.

Saturn

This occultation and can account for those sudden fits of depression and unhappiness that seem to occur without you having a real reason for them. The truth is that the occultation merely brings to the surface a problem that has been building up for a very long time. Sometimes this involves a father or father figure, or yourself in the guise of a father. Something may affect the fabric of your home or the goods that you have in it, but it's more likely to be connected to the people who are in and around your home. Your job, business, career or finances will endure setbacks.

Uranus

This signifies a sudden event. The problem is that if the occultation goes on for month after month, your life will continue to be rocked by a series of unpredictable events and occurrences. The fact that the moon is involved suggests that something odd will happen in your home or among family members, perhaps even to one family member after another. If you are in the throes of trying to move house or to refurbish one, there could be setbacks and upsets. This is probably not the right time (or even the right decade) in which to try to do things of this kind.

You may find yourself in a phase where one friend after another leaves your orbit, lets you down or hurts you in some way. Perhaps a group or even a club that is important to you becomes a problem. If you are into teaching or training, either as a tutor or as a student something could be wrong here.

Neptune

This can cause practical difficulties such as flooding or some other water-related problem that necessitates calling in plumbers and builders. Another more dangerous situation is that of a gas leak, so get all gas appliances checked out when Neptune is going through an occultation series.

You may suddenly lose interest in a creative venture or decide to give up studying an artistic or musical subject, or you may suddenly decide that you need to take up something of this kind as an antidote to the mundane nature of your daily life. It's just possible that you could find yourself a prey to spiritual or psychic events of the less pleasant kind, and you may even need to call in an exorcist!

If drink, drugs or some strange health problem or allergy is part of your life, expect these to become really troublesome now. This may, in effect, constitute a wake-up call to do something about your health and well-being.

Pluto

Fortunately, this planet is unlikely to follow the pattern of being occluded by the moon over a long period of time - and it's rare that it's ever involved in an occlusion. This is because Pluto has a more up and down orbit than any other planet and it's rarely on the ecliptic or in any place where it could suffer from repeated occultations by the moon. If it were, then this would signal a long period of destruction and devastation, followed by another long period of transformation.

Chiron
The ephemeris doesn't cover Chiron, and fortunately, this planetoid has an uneven orbit, so it probably doesn't get involved in many occultations. If it did, it would signal health problems or painful emotional situations.

Appendix

A Few Famous People

Here are the sun signs, moon signs, ascendants and nodes of a selection of celebrated personalities. Check out those that interest you, with the ideas in this and other books and see how well they fit.

Aries North Node - Libra South Node

Richard M Nixon
Sun Capricorn, fifth house.
Moon Aquarius, sixth house.
Ascendant Virgo
North node Aries, seventh house.
South node Libra, first house.

Sean Connery
Sun Virgo, seventh house.
Moon Virgo, eighth house.
Ascendant Capricorn
North node Aries, second house.
South node Libra, eighth house.

Meryl Streep
Sun Cancer, eleventh house.
Moon Taurus, tenth house.
Ascendant Leo.
North node Aries, tenth house, conjunct the midheaven.
South node Libra, fourth house, conjunct the nadir.
MC conjunctions often indicate fame and fortune.

Richard Branson
Sun Cancer, twelfth house, conjunct the ascendant.
Moon Virgo, second house.

Ascendant Cancer.
North node Aries, conjunct the midheaven.
South node Libra, conjunct the nadir.
MC conjunction indicates fame and fortune.

Taurus North Node - Scorpio South Node

Al Gore
Sun Aries, ninth house.
Moon Capricorn, fifth house.
Ascendant Leo.
North node Taurus, tenth house.
South node Scorpio, fourth house.

Grace Kelly
Sun Scorpio, first house.
Moon Pisces, fifth house.
Ascendant Scorpio.
North node Taurus, seventh house.
South node Scorpio, first house.

Prince Charles
Sun Scorpio, fourth house.
Moon Taurus, tenth house.
Ascendant Leo.
North node Taurus, tenth house.
South node Scorpio, fourth house.

Gemini North Node - Sagittarius South Node

Diane Keaton
Sun Capricorn, second house.
Moon Aquarius, eighth house.

Ascendant Scorpio.
North node Gemini, eighth house.
South node Sagittarius, second house.

George W. Bush
Sun Cancer, twelfth house.
Moon Libra, third house.
Ascendant Leo.
North node Gemini, eleventh house.
South node Sagittarius, fifth house.

Stephen Spielberg
Sun Sagittarius, sixth house.
Moon Scorpio, fifth house.
Ascendant Cancer.
North node Gemini, eleventh house.
South node Sagittarius, fifth house.

O J Simpson
Sun Cancer, eleventh house.
Moon Pisces, eighth house.
Ascendant Leo.
North node Gemini, tenth house.
South node Sagittarius, fourth house.

Cher
Sun Taurus, eleventh house.
Moon Capricorn, seventh house.
Ascendant Cancer.
North node Gemini, twelfth house.
South node Sagittarius, sixth house.

Cancer North Node - Capricorn South Node

Goldie Hawn
Sun Scorpio, eleventh house.
Moon Gemini, sixth house, conjunct the descendant.
Ascendant Sagittarius
North node Cancer, conjunct descendant.
South node Capricorn, conjunct ascendant.

Michael Douglas
Sun Cancer, tenth house.
Moon Capricorn, second house.
Ascendant Scorpio.
North node Cancer, eighth house.
South node Capricorn, second house.

Miles Davis
Sun Gemini, twelfth house.
Moon Scorpio, sixth house.
Ascendant Gemini.
North node Cancer, second house.
South node Capricorn, eighth house.

Leo North Node - Aquarius South Node

George Bush
Sun Gemini, tenth house.
Moon Libra, second house.
Ascendant Virgo.
North node Leo, twelfth house.
South node Aquarius, sixth house.

George Harrison

Sun Pisces, fourth house.
Moon Scorpio, twelfth house.
Ascendant Scorpio.
North node Leo, tenth house, conjunct the midheaven.
South node Aquarius, fourth house, conjunct the nadir.

Thrust into the limelight? MC conjunctions bring fame and fortune.

Virgo North Node - Pisces South Node

Prince Andrew

Sun Pisces, seventh house.
Moon Scorpio, fourth house.
Ascendant Leo.
North node Virgo, third house.
South node Pisces, ninth house.

The Late Princess Diana

Sun Cancer, seventh house.
Moon Aquarius, second house.
Ascendant Sagittarius.
North node Virgo, eighth house.
South node Pisces, second house.

Libra North Node - Aries South Node

Martin Sheen

Sun Leo, sixth house.
Moon Leo, sixth house, conjunct descendant.
Ascendant Aquarius.
North node Libra, eighth house.
South node Aries, second house.

Richard Burton
Sun Pisces, fifth house.
Moon Leo, ninth house.
Ascendant Scorpio.
North node Libra, twelfth house.
South node Aries, sixth house.

Scorpio North Node - Taurus South Node

Lily Tomlin
Sun Virgo, third house.
Moon Aries, tenth house.
Ascendant Cancer.
North node Scorpio, fourth house.
South node Taurus, tenth house.

Natalie Wood
Sun Cancer, tenth house.
Moon Taurus, seventh house.
Ascendant Libra.
North node Scorpio, second house.
South node Taurus, eighth house.

Sagittarius North Node - Gemini South Node

Nat King Cole
Sun Pisces, second house.
Moon Libra, eighth house.
Ascendant Capricorn.
North node Sagittarius, eleventh house.
South node Gemini, fifth house.

Capricorn North Node - Cancer South Node

Woody Allen
>Sun Sagittarius, fourth house.
>Moon Aquarius, sixth house.
>Ascendant Virgo.
>North node Capricorn, fifth house.
>South node Cancer, eleventh house.

John F Kennedy
>Sun Gemini, eighth house.
>Moon Virgo, eleventh house.
>Ascendant Libra.
>North node Capricorn, third house.
>South node Cancer, ninth house.

Monica Lewinski
>Sun Leo, conjunct midheaven.
>Moon Taurus, seventh house.
>Ascendant Libra.
>North node Capricorn, third house.
>South node Cancer, ninth house.

Aquarius North Node - Leo South Node

Elvis Presley
>Sun Capricorn, second house.
>Moon Pisces, third house.
>Ascendant Sagittarius.
>North node Aquarius, second house.
>South node Leo, eighth house.

Tony Blair
>Sun Taurus, twelfth house.

Moon Aquarius, tenth house, conjunct the midheaven.
Ascendant Gemini.
North node Aquarius, ninth house, conjunct the midheaven.
South node Leo, third, conjunct the nadir.
Unexpected fame and fortune due to the moon/node conjunction.

Pisces North Node - Virgo South Node

Princess Anne
Sun Leo, tenth house.
Moon Virgo, eleventh house.
Ascendant Libra.
North node Pisces, fifth house.
South node Virgo, eleventh house.

Elizabeth Taylor
Sun Pisces, third house.
Moon Scorpio, eleventh house.
Ascendant Sagittarius.
North node Pisces, third house.
South node Virgo, ninth house.

Michael Caine
Sun Pisces, ninth house, conjunct the midheaven.
Moon Libra, fourth house, conjunct the nadir.
Ascendant Gemini.
North node Pisces, ninth house.
South node Virgo, third house.

Catherine Zeta-Jones
Sun Libra, tenth house.
Moon Pisces, fourth house.
Ascendant Sagittarius.
North node Pisces, third house, conjunct the midheaven.

South node Virgo, ninth house, conjunct the nadir.
The north node close to the midheaven brings fame and fortune.

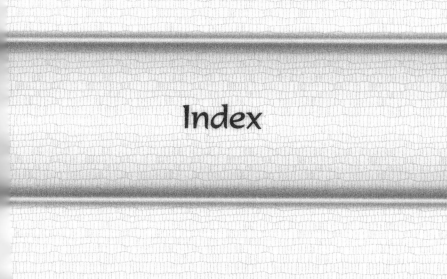

Index

Zambezi Publishing Ltd

We hope you have enjoyed reading this book. The Zambezi range of books includes titles by top level, internationally acknowledged authors on fresh, thought-provoking viewpoints in your favourite subjects. A common thread with all our books is the easy accessibility of content; we have no sleep-inducing tomes, just down-to-earth, easily digestible, credible books.

~~~~~

Please visit our website (www.zampub.com) to browse our full range of Lifestyle and Mind, Body & Spirit (MB&S) titles, and to discover what might spark your interest next...

~~~~~

For an equally absorbing range of non-MB&S titles and details of all our ebooks, visit our sister website, www.stelliumpub.com

Lightning Source UK Ltd.
Milton Keynes UK
UKOW06f1803191215

264997UK00011B/172/P